The Birth of the Anthropocene

The publisher gratefully acknowledges the generous support of Deborah & David Kirshman/Helzel Family Foundation, Judith & Kim Maxwell, and Barclay & Sharon Simpson as members of the Literati Circle of the University of California Press Foundation.

The Birth of the Anthropocene

JEREMY DAVIES

UNIVERSITY OF CALIFORNIA PRESS

University of California Press, one of the most
distinguished university presses in the United
States, enriches lives around the world by advancing
scholarship in the humanities, social sciences,
and natural sciences. Its activities are supported
by the UC Press Foundation and by philanthropic
contributions from individuals and institutions.
For more information, visit www.ucpress.edu.

University of California Press
Oakland, California

Library of Congress Cataloging-in-Publication Data

Names: Davies, Jeremy, 1983- author.
Title: The birth of the Anthropocene / Jeremy
Davies.
Description: Oakland, California : University of
California Press, [2016] | "2016 | Includes
bibliographical references and index.
Identifiers: LCCN 2015043076 (print) | LCCN
2016005793 (ebook) | ISBN 9780520289970 (cloth :
alk. paper) | ISBN 9780520289987 (pbk. : alk. paper) |
ISBN 9780520964334 (e)
Subjects: LCSH: Global environmental change. |
Nature—Effect of human beings on. | Environmental
geology. | Paleoecology—Holocene.
Classification: LCC GE149 .D38 2016 (print) |
LCC GE149 (ebook) | DDC 304.2—dc23
LC record available at http://lccn.loc.gov/
2015043076

Manufactured in the United States of America

25 24 23 22 21 20 19 18 17 16
10 9 8 7 6 5 4 3 2 1

In keeping with a commitment to support
environmentally responsible and sustainable
printing practices, UC Press has printed this book
on Natures Natural, a fiber that contains 30%
post-consumer waste and meets the minimum
requirements of ANSI/NISO Z39.48-1992 (R 1997)
(*Permanence of Paper*).

In memory of Jeanette Holt
1964–2014

CONTENTS

ACKNOWLEDGMENTS

I was able to write this book thanks to the generous study leave programs of the School of English and the Faculty of Arts at the University of Leeds. I spent some of my leave at the University of British Columbia, which gave me a perfect setting in which to write; my thanks to Alex Dick for his hospitality there. A remarkable group of people read drafts of the entire manuscript: Dipesh Chakrabarty, David Christian, Angela Clough, Ben Dibley, Alaric Hall, David Higgins, Daniel Lord Smail, Jan Zalasiewicz, and one other, anonymous reader for the University of California Press. Their expert revisions made the book much better than it would have been and saved me from many errors. I take full responsibility for the mistakes that no doubt remain. Anthony Carrigan, Alan Haywood, Jim Mussell, Stefan Skrimshire, and Graeme Swindles gave me valuable help and advice. Julia Banister kept telling me to make things clearer. Jamie Whyte drew the beautiful maps in chapters 4 and 5. Dore Brown, Bradley Depew, Bonita Hurd, and Niels Hooper have been an exemplary editorial team. I thank them all.

Introduction

This is a book about how to take the measure of a crisis. It is hard to grasp the scale of the modern environmental crisis, and part of the reason is that many things that had once seemed almost immutable are now changing rapidly.

The sea, for instance, is getting deeper. The world's oceans are likely to grow in height by between 40 and 120 centimeters before the end of the present century, letting them spill onto coastal land, where cities have always clustered. The cycle of the seasons is changing. The times are out of joint for plants like the early spider orchid, which has evolved to deceive mining bees into "pseudocopulation" as its only means of pollination: warmer springs mean that the bees emerge too early to be seduced by the flowers that depend upon them. Similar decouplings threaten many other lifecycles, like those of the birds who now hatch their eggs too late to catch the caterpillars that feed their young. Even the map of the world is being redrawn. The rivers that sustained the Aral Sea have been diverted for irrigation, shrinking it to barely a tenth of its former size. Sand and salt from the exposed lake bottom, mixed with pesticides, heavy metals, and defoliants, now blow onto the surrounding farmlands, making crop yields plunge and afflicting local farmers with asthma, tuberculosis, eye problems, typhoid, and cancer, and with kidney ailments from the saltiness of their drinking water.[1] Taken all together, this revolution that raises the oceans, reschedules the year, and turns water to land is bringing about a new epoch in the history of the world.

That last sentence might sound more declamatory than insightful, but in geology the word *epoch* has a specific technical meaning. A geological epoch is a midsize section of the planet's history. Students of the earth's biology and physical processes are now increasingly persuaded that the planetary system as a whole is undergoing an epoch-level transition. Earth's atmosphere, oceans, rocks, plants, and animals are experiencing changes great enough to mark the ending of one epoch and the beginning of another. The present environmental crisis is epochal in this particular, specialized sense. It is hard to comprehend its magnitude, but if we regard current environmental changes as the birth pangs of a new epoch, and if we give that epoch its place in geological time, in the long history of the earth itself, we might start to make sense of what we are facing. Recognizing what is now ending and what is beginning can help us respond to the predicament of living in the fissures between one epoch and another. The incipient new division of geological time has already been given a name: the Anthropocene. The idea of the Anthropocene epoch lets us understand the ecological crisis of the present day in the context of the distant past.

The central argument of this book is that the idea of the Anthropocene provides both a motive and a means for taking a very, very long view of the environmental crisis. It gives the ecological upheavals of the present day their proper place in the history of the planet. If you want to grasp the force, the scale, and the shape of the catastrophe as it unfolds, look for how it opens a fresh chapter in the long sequences of planetary time. To make sense of climate change, biodiversity loss, rain forest logging, and the rest, pay attention to how the current and imminent states of the world compare to those seen in the various epochs that went before.

If contemporary environmental changes add up to the birth of a new geological epoch, then earth scientists should ready themselves to adjust the geological timescale, the diagrammatic summary of the history of the planet upon which the whole science of geology rests. For now, the Anthropocene is not included on the official chart of the timescale that is maintained by its designated custodians, the International Commission on Stratigraphy. But a simplified and abbreviated version of that chart, with the Anthropocene added to it, would look like the diagram in figure 1.

Geological epochs such as the proposed Anthropocene are subsections of larger time units: periods, like the current Quaternary; eras; and ultimately eons. Epochs can themselves be subdivided into units called ages (not shown in this simplified diagram). All of these divisions and subdivisions come with fixed start dates and end dates, specified with greater or lesser margins of uncertainty according to the present state of geological knowledge. Evidently, when stratigraphers—experts in the physical sequences of rock strata upon which geological time sequences are built—postulate the beginning of a new epoch, they are making a quite specific claim. They envisage introducing one new piece, of a certain size and shape, into the carefully wrought mosaic of the geological timescale. The significance of the new interval, like that of all the older ones, would depend in large part on when it was said to have begun. Its hierarchical status, too, would matter greatly: to declare a new epoch would be a smaller step than creating an Anthropocene period, but an epoch would loom larger in geologic time than a mere Anthropocene age. So when it is used by stratigraphers, the word *Anthropocene* designates an interval that would occupy one particular place within the immense volume of geological time.

As yet the stratigraphers' debates about the Anthropocene, and the ins and outs of their conclusions, have never been examined at all closely from outside the tradition of the earth sciences. One of my aims in this book is to introduce other readers to the perspective on environmental history that has

EON	ERAS	PERIODS	EPOCHS
Phanerozoic since 541 million years ago	Cenozoic since 66 mya	Quaternary since 2.58 mya	Anthropocene
			Holocene
			Pleistocene
		Neogene 23–2.58 mya	Pliocene
			Miocene
		Paleogene 66–23 mya	Oligocene
			Eocene
			Paleocene
	Mesozoic 252–66 mya	Cretaceous	Subdivision into 31 epochs
		Jurassic	
		Triassic	
	Paleozoic 541–252 mya	Permian	
		Carboniferous	
		Devonian	
		Silurian	
		Ordovician	
		Cambrian	

Figure 1. Phanerozoic eon.

emerged from those debates. That perspective—which begins with an assessment of the geological traces that the last few centuries will leave behind in the distant future—has the potential to be enlightening for anyone concerned about the environment, not just geologists. But this book also has a much larger aim. I argue that the stratigraphers' version of the Anthropocene can yield a new way of understanding and responding to the modern ecological catastrophe. The catastrophe is so far-reaching that it cannot really be understood without setting it in the context of geologic time. That means that the long view provided by geology can change the basics of environmental politics for the better. The Anthropocene of the stratigraphers opens a window onto the geological past, and the politics of the environment can be put on a new footing by the stratigraphic vision of the new epoch.

With contemporary politics in mind, the most immediate and most telling point of comparison for the Anthropocene is the Holocene epoch, the 11,700-year span of time that in the established version of the geological timescale still continues to the present day. I believe that in order to make sense of this comparison between the Holocene and the Anthropocene we will also need to look much further back into the geological past, where monsters abound. But the first crucial point is that introducing an Anthropocene epoch to the geological timescale (and placing its starting point somewhere in the last few centuries) would mean declaring that the Holocene is now arriving at its end. This book, then, will eventually be just as much about the terminal crisis of the Holocene as it is about the birth pangs of the Anthropocene, or rather, I emphasize that those two things are one and the same. The Holocene matters because it is the only geological epoch so far in which there have been symphony orchestras and hypodermic needles, moon landings and gender equality laws, patisseries, microbreweries, and universal suffrage—or, to put it plainly, the agricultural civilizations that eventually made all of those things possible. With its demise, the civilized rights and pleasures previously confined to the Holocene will have to negotiate radically changed ecological conditions if they are to endure, let alone if they are to be extended more generously to more people. That is the political problem of the Anthropocene.

It is always intellectually stimulating to find that you are positioned in the interstices of two different worlds. The idea of the Anthropocene makes this state of being in between epochs the starting point for political thinking. In the last chapter of this book, and in the conclusion, I argue that environmentalists should think of themselves as being caught up in the transition between two geological intervals, and that the goal of environmentalism should be to negotiate a way through this transition. That means demoting the ideal of "sustainability" from its status as the greens' highest objective. Instead, environmental movements will need to be concerned above all with environmental injustice and with fostering ecological pluralism and complexity in the face of the simplifying tendencies of the Holocene's final phase. The birth of the Anthropocene should be attended by vigilant resistance against the searing away of multifaceted socioecological systems and their replacement by vulnerable, saturated monocultures. Or to put it more positively, the jerky crossing between epochs can be cushioned by upholding states of life—both ecosystems and human societies—that are variegated, intricate, and plural, ones in which lively forces of all kinds contend with and interweave with one another.

The word *Anthropocene* is descended from the Greek ἄνθρωπος (*anthropos*), meaning either "man" or "human." It is a recent addition to the vocabulary of environmental politics: it was coined, or at least it came to something like widespread notice, only at the end of the twentieth century. But since then it has prospered in a remarkable way. In some academic circles it has lately become a much-used and fashionable term. In the most advanced circles of all it has already gone on to the next stage and is considered rather worn-out and déclassé. Among both the enthusiasts and the skeptics the word has been tossed into debate much more frequently than it has been explained or defined. More often than not, it has been used without the intention of any very specific allusion to the work of the stratigraphers that provides its significance in the context of this book. That's fine, of course. There is no reason why the word should not be used in a whole range of diverse, contested, and even incompatible ways. For the sake of clarity, however, I would like to set out, before going any further, some of the things that "the Anthropocene" will not mean in the pages that follow.

Firstly, the Anthropocene, in this book, is not the name of a fall from Eden. It does not describe the period in which human acts have brought about the end of nature by pollution and despoliation: it is not a rhetorical device to make clear the extent of human depravity. It follows that the Anthropocene is not the kind of thing that it is possible to "mitigate," like an oil spill. Secondly, and conversely, the Anthropocene is not a breakthrough from tedious natural stasis. It is not the transcendence of the earth's old limits, the sundering of its chains. It does not stand against all previous epochs and periods, looking glamorous and disreputable where they were worthy and dull. It is one epoch among many on the same footing, rather than one-half of the earth's history.

Thirdly, despite its name, the Anthropocene is not an anthropocentric concept. The epoch does not get its name because nature is now completely subordinated to human agency, as if clouds now form and swallows now fly only after getting permission from human beings. The name suits it because human societies exert a novel and distinctive degree of sway in the physical world, but other creatures still continue independently to exert their own powers and to pursue their own interests in this new field of action. Humanity is not at the center of the picture of the Anthropocene, opposing, by its powers of mind, the passive matter that encircles it. Instead, human societies are themselves constructed from a web of relationships between human beings, nonhuman animals, plants, metals, and so on. Nor, fourthly, is the Anthropocene a concept that reduces humankind to an undifferentiated mass. I will return—at some length—to that point. To say that the earth is undergoing an epoch-level physical transition, in which the activities of sundry groups of humans are playing key roles, does not imply in the least that all human beings have thus far acted in unison, or that they are all collectively responsible for the new state of affairs.

Finally, in arguing for the importance of looking at the environmental crisis in the context of geological time, I am not at all advocating a distanced,

Olympian perspective on the human condition. Even though the requisite context is prodigiously broad, paying attention to it does not mean rising above the present emergency in a spirit of enlightened impartiality. It does not mean drawing a contrast between the mere fleeting turbulence of humankind's concerns and the eternal currents of the great stream of life, and then looking with cool equanimity to the remote past and future where civilization is as nothing. In fact, it can mean exactly the opposite. Against the facile amorality of the truism that nature will not miss humankind after humans' inevitable demise, the idea of the Anthropocene may yield above all a sense of locatedness in time, a sense of being caught in one particular historical moment.

In a word: no more clean breaks that put humans on one side and nature on the other and, thereby, merge each antagonist into a uniform blob. I argue in this book that the birth of the Anthropocene does something quite different. It redistributes agencies, reconfigures systems, and reorders the loops of consequence and assimilation out of which the workings of the earth are made. The transition from one epoch to another is a generalized disruption, a drawing up of new accounts.

The opponents of the Anthropocene (of whom there are already many) often worry that the new word implies a bleak and narrow-minded picture of the world. In that picture, the planet has become a merely artificial construct, passively molded by human activity, and the best remaining hope for humanity is to allow a scientific elite to administer global affairs from the top down, so that natural resources may be exploited in the most efficient way and affluent consumer lifestyles may be kept afloat for as long as possible. I share those critics' dislike of such a scenario. But this book puts forward a very different world-picture. Here, the world is seen as characteristically full of devious chains of cause and effect; of intricate braids that link economies to ocean currents and ecosystems to plate tectonics; and of what climatologists call "teleconnections," far-distant perturbations that prove to be coupled by hidden bonds—although here the teleconnections can take the form of trade routes and cash flows as well as seesaws in atmospheric pressure. Feedback circuits let subtle evolutionary and chemical modifications have worldwide effects. Human societies exert their influence on the planet and so provoke the latest twist in a chancy, surprise-filled geological history.

The recognition that the world is in the midst of an epoch-level transition is of a piece with the general tenor of earth science research over the last forty years. During that time, a conceptual framework usually called *neo-catastrophism* has come to the forefront of the earth sciences. I propose in this book that the idea of the Anthropocene should be seen as another product of that neocatastrophist turn. Neocatastrophism has enlivened modern geo-science by dispatching the belief that the planet took on its current shape only through the gradual and continuous operation of familiar processes like erosion and sediment buildup. The new geology lets into the picture abrupt die-offs and bursts of species formation, climatic and geomorphological upheavals, and high-speed collisions with extraterrestrial bodies. Bit by bit, the life of the earth before human civilization has come to look ever more dramatic and incident-packed. There was no stately, teleological progress toward the arrival of humans. Instead, the story has been full of sharp twists and transformations. Built into the earth system are a multitude of concate-nated feedback mechanisms. These feedback mechanisms have repeatedly amplified even comparatively small initial changes in unpredictable ways, making nonhuman history as contingent and chaotic as the history of king-doms and empires.

This new understanding of the earth system has greatly influenced cli-mate scientists, for instance. As they keep struggling to explain, the reason to be concerned about global warming is not that the composition of the atmosphere is now altering rapidly for the first time ever, or that it is dis-rupting the eternal harmony of the climate system to frighteningly unknow-able effect. On the contrary, it is that the atmosphere and the climate have changed swiftly and mightily from time to time in the past. These changes have tended to bring with them a new configuration of living things, one that—however fine in itself—has been to the old one like a conquering army to a fallen city. That ominous historical record is the reason why contemporary perturbations to the climate system are at the heart of the dangers posed by the birth of the Anthropocene.

Neocatastrophism has introduced us to a whole list of geophysical forces—asteroids, ocean currents, volcanoes, and the like—that, under the right circumstances, can suddenly come to exert a much greater and more destabilizing influence than usual on the workings of the earth system. The idea of the Anthropocene, as I want to construe it, simply adds human agency to that list. The Anthropocene gets its name from humans, the *anthropos*, because its distinguishing characteristic (for now) is the dramatic influence that human societies are having on the physical world. It is not the case that human interventions in the earth's organic makeup, or in the processes governing its soil or water or atmospheric cycles, are still dwarfed by any mightier forces that transcend humankind's paltry strength. Far from it. Human societies are now among the most powerful of the ecological forces that operate on, above, and below the surface of the earth.

In this light, perhaps the most incisive account of the new epoch so far has come not from a scientist or a campaigner but from a poet, the Canadian Don McKay. McKay's rich body of work has been characterized most of all by his interests as a birder. In his two most recent collections, however—*Strike/Slip* and *Paradoxides*—his line of sight has turned lower and slower. Geology has become the keynote of his poetry, which has hunkered down among fossils, rocks, and tales drawn from deep time (that is, by analogy with "deep space," the abyss of time that stretches back from a few thousand years ago to the beginnings of the earth). McKay has written poems about hexagons of quartz that formed long before the first mathematics, about stumbling across a trilobite on the shore of the North Atlantic, about the imponderability of hundred-million-year timescales and the wearing away of mountains. In a lecture in 2008 he reflected on the uses of the Anthropocene. "All poets take naming seriously," he observed, and for him, giving a name to the Anthropocene creates for us "an entry point into deep time." The preceding geological epochs seem to run backward from this new one "like rungs on a ladder" descending within a few steps into a time before human existence. With a quantity of blunt sarcasm, McKay lays out what seems to me the profoundest significance of the birth of the Anthropocene:

If we think of ourselves as living in the Anthropocene Epoch, we realign our notion of temporal dwelling. Generally, time is viewed in relation to humanity's place in it, and consists of a present, where we live, and a recent past called history, which is felt to be important for informing the present and helping us understand ourselves better. When we speak of the past with reverence or chagrin, it is this shallow past we mean. Before history there is a vague distant past called prehistory, comprised of a jumble of relics and catastrophes, dinosaur bones mixed with clovis points, missing links, Lucy and The Flintstones cohabiting in the caves of Lascaux, Australopithecus confused with archaeopteryx, and the whole *mélange* construed as a sort of amniotic stew from which we, the Master Species, miraculously emerged. The name "Anthropocene," paradoxically enough, puts a crimp in this anthropocentrism, making the present a temporal unit among other epochs, periods and eras. . . . On the one hand, we lose our special status as Master Species; on the other, we become members of deep time, along with trilobites and Ediacaran organisms. We gain the gift of de-familiarization, becoming other to ourselves, one expression of the ever-evolving planet. Inhabiting deep time imaginatively, we give up mastery and gain mutuality.[2]

The Anthropocene sweeps humankind into the turbulent flow of geohistory. It announces a new intimacy with the older rungs on the ladder. "We"— and there will be much need to examine the implications of that collective pronoun—join the trilobites as actors in the long drama of life on earth: as another planetary force exerting its powers of survival and transformation. More than anything else, the Anthropocene is a way of thinking with deep time.

The best guides to this wild drama of deep time are the most fastidious and bookkeeping of figures in the profession of geology: the stratigraphers, who devote their labors to the precise demarcation and time-tabling of the deposition of rock layers all around the world. They have sought to measure the nascent epoch against the strict and cautious criteria that they have established for the formalization of geological intervals. The willingness of some stratigraphers to take on that task has given rise to the most vivid, the most radical, and the most disconcerting of all conceptions of the Anthropocene as it comes into being. It is their Anthropocene, a brand-new epoch to join the dozens that preceded it, that is my subject here.

In the first chapter that follows, I draw attention to the place of deep time in contemporary environmental news reporting. News stories often describe modern-day environmental changes as being unprecedented for thousands or even millions of years. That sounds not only sinister but also potentially confusing to anyone who is not an expert in earth history—a category that includes the great majority of people who are concerned about environmental issues. I criticize some unhelpful ways of imagining deep time, and describe how an alternative, geological perspective has grown up since the late eighteenth century. I also explore the question of just how much influence human societies currently have over the workings of the living planet. The idea of the Anthropocene itself enters the scene in chapter 2. Since the earth system scientist Paul Crutzen coined the word at the end of the twentieth century, its use has spread ever more widely. I trace the most important of those uses, and the backlash against the term that has developed in the last few years, before arguing that at least some versions of the Anthropocene are not guilty of the charges—of anthropocentrism and antidemocratic arrogance—that have been brought against it.

Chapter 3 looks in detail at just one way of thinking about the Anthropocene. This is the stratigraphic version of the term, the one that takes it literally as a potential new addition to the geological timescale. I explore how the implied relationship between the Anthropocene and the *anthropos* changes when the word is taken in a stratigraphic sense, and I describe the thought experiment that underpins the stratigraphers' approach: if alien geologists were to arrive on the earth in a hundred million years' time, what fossilized traces of present-day events would they find? I spend a long while on the seemingly hairsplitting question of when exactly the new epoch should be said to begin, because that question proves to be a way of addressing the crucial issue of how geological designations can reflect the environmental history of the world over the last several centuries.

Those first three chapters describe how the idea of the Anthropocene can open up a window on geological time. The final two chapters offer a look through that window. The main part of each one is a broad-brush narrative time line. Chapter 4 surveys the Phanerozoic eon, the 541-million-year interval within which the Anthropocene ultimately belongs, and chapter 5 surveys the Holocene epoch, the Anthropocene's immediate predecessor. The aim of those narratives is to give life and significance to the geological timescales that are the necessary points of reference for the new epoch, timescales that might otherwise look blankly intimidating to many environmentally conscious people who do not happen to be professional geologists. Along the way, chapter 4 considers the place of *Homo sapiens* in deep time, and chapter 5 considers the place of civilization in the period since the end of the last ice age. In the conclusion, I tease out the political implications of the idea of the Anthropocene epoch. It can be both a polemical slogan and a conceptual basis for environmental politics. Talk of sustainability, and of respecting the ecological limits to growth, tends to imply a forlorn attempt to escape from temporal constraints. In contrast, a stratigraphic perspective makes the specifics of the present crisis the point of origin for environmentalism. A politics grounded on the attempt to dwell within and to shape the terminal crisis of the Holocene epoch would be transnational in its spirit and committed to analyzing the inequalities of power that often trigger environmental catastrophe. Its aim would be to foster a raucously democratic pluralism in the ecosystems of the birth of the Anthropocene.

CHAPTER ONE

Living in Deep Time

The Anthropocene epoch offers a way to understand the present environmental crisis in the context of deep time, the realms of the distant geological past. And as a strange recent tendency in environmental news reporting shows, current generations are being plunged into deep time, like it or not, by the once-in-a-million-year environmental changes that are taking place around them. Climate change deniers share with some well-intentioned environmentalists a damaging and unrealistic view of the planet's deep past as an essentially static state of affairs. But since the end of the eighteenth century the sciences of the earth have developed a very different way of looking at the distant past, a perspective that has grown ever more clearly defined thanks to some major developments in geological thought during the last few decades. In this alternative view, geological time is historical through and through. Tracing its story reveals a dynamic narrative of floods, climate changes, and unpredictable evolutionary development. The birth of the Anthropocene epoch is best seen as the latest turning point within the swirling history of deep time. But if the story of the earth has always been so lively, one might wonder whether present-day change is in fact all that noticeable in the grand scheme of things. What, then, is the real scale of human-induced changes to the earth's systems as a whole?

THE LONG MOUNTAIN

Early in May 2013, at an observatory on the black volcanic slopes of Mauna Loa, Hawaii, the daily average concentration of carbon dioxide measured in the atmosphere rose above 400 parts in every million. The level declined by some 7 parts per million over the next few months as CO_2 was drawn from the sky by summer vegetation across the Northern Hemisphere, before it began to rise again in the autumn. The following year, the 400 ppm threshold was crossed in March. A year later, it was crossed in January.

The air of Mauna Loa, the "long mountain" that ascends from the middle of the Pacific, has long been closely monitored. The mountain's remoteness, and the lunar barrenness of its upper slopes, mean that its bright clean air can serve to indicate the state of the whole planet's atmosphere. And because the chemical composition of the atmosphere has been an intensely political issue ever since the beginning of public concern about greenhouse gases in the late 1980s, the first crossing of the 400 ppm limit at the Mauna Loa station was widely reported. The rapidly increasing carbon dioxide level was understood to be the consequence of human activity, and to be cause for concern about the changing state of the climate. The newspapers that reported the story also felt the need to supply some historical context. As recently as the middle of the eighteenth century, journalists explained, CO_2 concentrations stood at around 280 ppm. Thus, given that no other factors can plausibly explain the increase, three-tenths of the current CO_2 level is attributable to the development of industrial society since the late eighteenth century.

In the complex field of climate science, what story could be clearer than this? After all, explanations of the contemporary world that look back as far as the late eighteenth century are perfectly familiar. That is most obviously and especially the case in the United States, where two documents written shortly before 1800, the original Constitution of the United States and the Bill of Rights, still frame a remarkable proportion of everyday political discussion. It is also true more widely, however. The last decades of the eighteenth century were a formative period for European colonial expansion, so the influence of that era can still be seen in the basic shape of the modern world, its unequal distribution of wealth and poverty. Moreover, the period of the Industrial Revolution also witnessed the French Revolution, the founding event of the modern liberal state. Asked about its impact in the 1970s, the Chinese statesman Zhou Enlai is often—although perhaps mistakenly—said to have replied, "It is too soon to tell." Authentic or not, Zhou's aphorism is admired as a telling expression of a plausible idea: that the impacts of the French Revolution are still playing out, and that contemporary politics still takes place partly in the shadow of 1789. But what if Zhou had said the same thing about (for instance) the formation of the Isthmus of Panama—the clasping of hands between North and South America, which divided the Pacific from the Atlantic three million years ago? What sort of political event, if any, might realistically be placed in a framework that stretches back not hundreds of years, but millions?

That question arises because the newspapers that reported on the phenomenon at Mauna Loa did not look back only as far as the eighteenth century. The journalists writing up the story evidently felt that their readers would be poorly informed if they were confined to such a short-term perspective. What was so special, after all, about the carbon dioxide levels of the mid-eighteenth century, just before the rise of industrialism? To explain that, the *New York Times* broadened its purview spectacularly. "For the entire period of human civilization, roughly 8,000 years, the carbon dioxide level was relatively stable near [280 ppm]." No doubt many of the *Times'* readers in May 2013 felt that, as conscientious modern citizens, they should be able to appreciate the significance of the climate change story on the front page of their daily newspaper. But it seemed as if in order to manage that, they would need to take on board not a mere couple of centuries of historical background but eight thousand years. Or rather, even doing that would get them no more than a hundredth of the way to understanding the story.

As the *Times* coolly told them: "From studying air bubbles trapped in Antarctic ice, scientists know that going back 800,000 years, the carbon dioxide level oscillated in a tight band, from about 180 parts per million in the depths of ice ages to about 280 during the warm periods between." Ice ages, in the plural! The mountaintop sages of Mauna Loa began to sound like those Hindu scholars who reflect on the hundreds of thousands of solar years that make up a yuga, each one a part of the mahā-yuga cycles that form one seventy-first part of a twenty-ninth of a day in the life of Brahmā. But the *Times'* reporter went further still. He finally set the morning's news from the Pacific in its proper context when he observed that "the last time the carbon dioxide level was this high was at least three million years ago, during an epoch called the Pliocene."[1]

The *New York Times* was not alone. In Britain, the *Guardian* wrote up the story with an explanation that the new CO_2 level had "not been seen on Earth for 3–5 million years, [since] a period called the Pliocene." Brazil's *O Globo* noted that carbon dioxide had not reached the "marca símbolo" of 400 ppm for "3,2 milhões de anos." In France, *Le Nouvel Observateur* reported the upper figure: it was perhaps "cinq millions d'années" since "l'atmosphère terrestre" had contained so much carbon dioxide. Again and again, a story about rising carbon dioxide levels in the atmosphere prompted invocations of the ancient past. Why? What was at stake in explaining one very small number—400 parts per million, or 0.04 percent—by invoking a second very large number, five million years? Why was a period of time usually left peacefully to geologists, paleontologists, and evolutionary biologists suddenly everybody's concern?

The journalists were surely right to think that the significance of the 400 ppm concentration could not be grasped without making some reference to the last time CO_2 levels were so high. This pressingly topical issue, to which politicians and pundits were prompt to respond, really did demand to be juxtaposed with the deep geological past. And more strangely still, the breaching of the 400 ppm threshold was far from alone in this respect. Any number of recent environmental changes, familiar to anyone who reads the papers, exhibit just the same doubleness. On the one hand, present-day political salience. On the other, legibility only through deep time. Talking about the current environmental crisis seems to mean that one also needs to talk about very distant seasons in the history of the earth.

Stories about melting glaciers come with references to when the world was last free of ice sheets, tens of millions of years ago. A report that temperatures in the Arctic are at their highest for at least forty-four thousand years becomes headline news. Disputes about government agencies' handling of floods or forest fires are framed by quotations from experts who describe how rivers shift their courses back and forth over thousands of years, or how certain species of woodpecker have evolved over millions of years to feed on the grubs that colonize burnt trees. Conservationists argue that the baselines for what constitute fully functioning ecosystems may need to be set at tens of thousands of years ago, before most large mammals were driven to extinction. Campaigners against global warming describe it as madness to burn up within a few decades coal and oil deposits that accumulated over many millions of years. Newspaper features about biodiversity loss are given urgency by the suggestion that the world may be starting to experience only the sixth mass extinction of the past half a billion years. The most seemingly transient phenomena can turn the attention of a concerned public to times long ago, as with the news of a study showing that the Roaring Forties, the persistent westerly winds in the Southern Hemisphere, are blowing more strongly and on a more southerly track than at any time in at least the last millennium.

This tendency is probably obvious to anyone who pays even casual attention to news stories about environmental issues. Nonetheless, it is easy to overlook just how noteworthy the tendency is. Individual references to deep time in environmental reportage often appear incidental or ad hoc. They reveal their real significance only because of how frequently they recur. Taken as a whole, these opportunistic media allusions make a crucial point. That is: in order to understand the current environmental crisis you have to think about very long ago. From year to year, and from decade to decade, the world of the early twenty-first century is undergoing changes that can be grasped only by switching to timescales of tens of thousands or even millions of years. Facts that politicians and pressure groups are prone to argue about, to assign blame for, and to promise their electorates or their memberships to ameliorate—contemporary political facts, in other words—need to be explained by referring to eras long before any such thing as politics even existed. Climate change, biodiversity loss, chemical pollution, and so on have made journalists talking to the public invoke geological time spans as casually as if they were paleontologists engaged in conversation with glaciologists.

That poses a problem, surely. The environmental catastrophe has politicized deep time. So how are people who care about the environment, but who are neither paleontologists nor glaciologists, supposed to deal with these vast expanses of history? How can they understand them, imagine them, or make sense of day-to-day environmental changes that are placed in this startling context? If we read that the federal minimum wage in the United States has declined to a real-terms level last seen in the 1950s, or that the richest 1 percent of Americans and Europeans are well on their way to securing their largest share of national wealth since before the First World War—comparisons that have the same structure as the Mauna Loa report—it is relatively easy to see the point that is being made. By contrast, the references to deep time bandied about in environmental news reporting are likely to be confusing and instantly forgettable for noninitiates. As one professor of geography wearily remarks, "It is common when asking new undergraduates about periods of past time when things may have happened . . . to find a random selection of answers that fails to differentiate between hundreds, thousands, hundreds of thousands and even millions of years."[2]

The single most memorable date in the ancient past—the equivalent, for the British, of the Battle of Hastings in 1066—is probably sixty-six million years ago. It was then that the terrestrial dinosaurs were eradicated by a comet or asteroid that struck the earth off what is now the coast of Mexico. With a few possible exceptions like that, there is no particular reason why ten million years ago should summon up mental images in the minds of nonspecialists that are very different from a hundred million years ago, or one million. If the last of those dates stands out, it might be only because of Hammer Films' *One Million Years B.C.*, with its stop-motion dinosaurs lurching toward underdressed cavewomen.

News reporters can mitigate the problem with nutshell explanations of the dates that they discuss. The *New York Times'* splash on the CO_2 levels at Mauna Loa included an explanation that three million years ago the climate "was far warmer than today, the world's ice caps were smaller, and the sea level might have been as much as 60 or 80 feet higher." That sort of gloss certainly helps, but only as a sort of decontextualized snapshot. Carbon dioxide concentrations and sea levels evidently do not correlate perfectly, and without a continuous narrative to hold on to, mapping the rise and fall of CO_2 and the oceans step by step over all this time, even the facts that the *Times* recorded might slip out of one's grasp. It could have been thirty million years ago that the earth was so warm and its air held so much carbon. It could have been a mere three hundred thousand. It's easy to forget.

One might object that something relatively similar is true of many news stories besides the one from Mauna Loa. Leafing through the newspapers, many of us, no doubt, would like to have a better grasp of the historical background to current affairs of all kinds, not just environmental ones. But those other stories are never quite analogous to the ones about the environment. Reading a headline about sectarian clashes in Northern Ireland, you might justifiably be far from certain about the year of the battle commemorated by the Orangemen's provocative marches on the twelfth of July, and only be able to hazard an estimate that it took place three or four hundred years ago. Still, no one thinks that the Battle of the Boyne took place thirty years ago, or three thousand. On the business pages, by contrast, misremembering figures by an order of magnitude is certainly possible: when AIG was bailed out during the crisis of 2008, did it cost billions, tens of billions, or hundreds of billions of dollars? But those fantasy numbers of the financial system operate by their own rules, and when they grow unmanageably large they become just one more example of stock markets' many arcana, not a main impediment to grasping the news of the latest crash or takeover.

The science pages might direct your attention back to the very origins of time, and describe the latest research into the Big Bang itself. But in that case the astrophysicists' conclusions (as opposed to the question of how their laboratory is funded) hardly sound like a political topic. Back on the news pages, demagogues on the ethnic-nationalist fringe assert one people's exclusive right to a territory on the basis that they have always been there. But even the most rabid and fantastical among them claim a tradition of ownership that goes back only a few thousand years at most. In the lifestyle section, you might read about the latest fad in dieting—paleo dieting—based on a theory about what "our hunter-gatherer ancestors" used to eat. But moralizing the distant past in that way (as evolutionary pop psychology also tends to do) is not quite the same as asserting that it has a pressing *political* relevance.

In short, politicizing deep time is a habit peculiar to environmentalism. Ecological politics struggles with the difficulty of imagining the distant past. Efforts to fix geological time in familiar, tangible terms only make it even stranger. At an inch to a year, the age of the earth is nearly equivalent to three circles round the equator. Or as one famously vivid illustration has it: "Consider the earth's history as the old measure of the English yard, the distance from the king's nose to the tip of his outstretched hand. One stroke of a nail file on his middle finger erases human history."[3] Certainly, the geological timescale can be learned: geology students need to do so in order to pass their exams. But it is hardly reasonable to make that memorization exercise a requirement for ecological good citizenship. Everybody is already overburdened with the weight of information available to them about the state of the planet. What is needed instead is some plausible way of coming to terms with the earth's bewildering antiquity, now that climate change and species loss have forced the subject forward into public attention.

TWO VERSIONS OF DEEP TIME

Something else makes it doubly urgent to find a rational way of setting environmental change in its deep-time context. Many previous allusions to the distant past in the context of ecological politics have been the very opposite of enlightening. Invocations of deep time are extremely common among those voluble bores who hold that climate science is a plot to improve researchers' access to funding grants or to impose a Marxist world government. The invaluable Skeptical Science website maintains a list of the most frequently exploited climate myths, and at the top of the list is a claim about deep time. That is: the climate is always changing, so climate change is not to be feared. "Climate has always varied; it is a special sort of narcissism to believe that only the recent climate is perfect," as the most highbrow of the global warming conspiracy theorists sneers.[4] Talk-show paleoclimatologists remark that the planet was warmer than today a thousand years ago (untrue),

8000 years ago (possibly true), 125,000 years ago (true, for now), indeed for the great majority of the time since the extinction of the dinosaurs (true by a large margin). Absurd, then, to worry about recent decades' correction back toward the long-term mean, given that deep time is full of warmer periods. This argument sometimes leads to the simple logical non sequitur that because past climate change was not caused by humans, neither is that of the present day. But the argument can also take a somewhat subtler route.

In this line of thinking, climate change is natural whatever its cause— aren't humans, too, a part of nature?—so there is no reason to worry about what the temperature is during any given period. Nothing natural is alien to the conspiracy theorists. Any climate recorded in the distant past is equally welcome in the future; the "alarmists" are only exploiting a superstitious fear of change. What underlies this way of thinking is a pretension to a standpoint of Olympian impartiality. To a totally disinterested observer, the climate regimen that humans have experienced for the past few thousand years is neither more nor less desirable than some tremendously hotter and wetter one. In those much hotter conditions, agriculture would fail and cities would drown, but palms and giant ferns would thrive. The world would be "lush" and "verdant." Objectively speaking, the earth would remain intact, no matter what the human death toll along the way. For the conspiracy theorists, then, the geological past serves as a universal stamp of validation, as a limitless repository of the natural.

What's more concerning is that a remarkably similar attitude to deep time can be seen among some well-meaning environmentalists. The use of deep time is often a way in which their beliefs coincide with the stratagems of their most cynical or most deluded opponents. Take the environmental thinker who has berated politicians for paying attention only to "desperately trivial twinklings of time" like the years and decades ahead. Colin Tudge writes, "It is impossible to contemplate the environment unless we think as a matter of course in very long periods of time indeed. In fact . . . *we cannot*

Living in Deep Time

claim to take the environment seriously until we acknowledge that a million years is a proper unit of political time."[5] That is bound to seem hopelessly intimidating, unless it just sounds pretentious. If everyone who is so trivial as to worry about what the next decade has in store is letting the side down, then the only people who take the environment seriously are a handful of well-educated contemplatives, at once high-minded and perfectly ineffectual. To be a good environmentalist, Tudge believes, means adopting a perspective that is no less Olympian than the one advocated by the global warming paranoiacs.

The same applies to that genre of green aphorisms in which the scale of the earth's history is used to generate eye-catching observations about the relative suddenness of modern environmental change. The wilderness guru David Brower used to travel the world giving a speech in which he condensed the story of the planet into the biblical six days of creation. Life began at noon on Tuesday, and "the beautiful, organic wholeness of it" developed over the next four days, he would say. But on Saturday night, "at one-fortieth of a second before midnight, the Industrial Revolution began. We are surrounded with people who think that what we have been doing for that one-fortieth of a second can go on indefinitely. They are considered normal, but they are stark, raving mad."[6]

Brower's speech—his "sermon"—was famous in its era and milieu, and many variations on his theme have been designed, circulated, and retweeted ever since. But the attention garnered by this half-truth comes at an indefensible cost. Brower's story seeks to bring confusion rather than clarity, to sweep away understanding in a moment of panicked sublimity. In his sermon the Industrial Revolution becomes the only real event in the history of the planet, on the model of a fall from paradise. The last two hundred years are reduced to a single, momentary blitz of criminal stupidity, a view that is mirrored and enabled by reducing what went before to a nearly infinite period of tedium. Humankind, busy and dynamic, is set in opposition to the prehuman world, taken as passive and static. That separation could scarcely be more distant from an ecological perspective whereby all human life is enmeshed within a web of planetary forces. Brower's sermon rested on the hope that a working environmental movement could grow from a feeling of bewilderment, from a startled sense that humans do not really belong on the planet, rather than from informed ecological citizenship.

To be clear, there is no reason to hold back from angry denunciations of ecocide, or from gestures of mourning for what has been lost: both have always been important to the environmental tradition. Equally, there need be nothing disheartening about an awestruck recognition of the antiquity of the earth. Fusing the two together, however, means presenting recent environmental change according to a stupefying timescale that denies it any meaning except as an emblem of the infinite culpability of humankind.

Deep time forms a single, beautiful organic whole both for David Brower and for the global warming conspiracy theorists. According to their shared understanding of the geologic past, all changes in earth's history before those of the last few centuries were equally natural, and for that reason the world was until then essentially unchanging—that is, not really subject to history at all. Perpetual but undifferentiated change throughout the life of the planet blurs into an effective stasis that is simply the natural state of affairs. If deep time is imagined as a single, homogenous mass, then with only the slightest change in perspective it can seem to justify either a primitivist excoriation of the whole industrial era, or an impatient dismissal of any possible ill effects from it. There must be better ways than this to imagine the distant past.

The sciences of the earth invite us to think of geologic time in a different way: as a drama without any preestablished outcomes. It is not an arena in which stable natural processes endlessly reproduce themselves, but a field of action dense with contingent successes and catastrophes. A *geohistory*. Only partly by coincidence, this is a way of thinking that was first documented at the very time and place of the start of the Industrial Revolution, at one proposed point of origin of the Anthropocene itself. Britain and France in the late eighteenth and early nineteenth centuries saw the emergence of what Roy Porter has called a "geological way of seeing." A new science, geology, arose from the idea that by examining rock strata one could discern both an irreducibly complex historicity and something, well, "spiritual." As Porter describes it, the geological way of seeing was an "analysis of the Earth in terms of great antiquity, the majesty of slow and profound process, the investigation of subterranean depths"; it meant that "to read the story of the strata was to read an autobiography of great revolutions, decay and restoration, the struggle of titanic Earth forces."[7] This was a perception that belonged in part to its historical moment, colored by Romantic landscape aesthetics and an unprecedented revolutionary impulse across Europe and America. It had, however, a permanently lasting effect. In the words of the preeminent authority on the subject, what took place was "the progressive transformation of the scientific study of the earth by the injection of historical ways of interpreting what can be observed: the earth, and by extension the natural world as a whole, came to be seen as having their own *histories*."[8]

The decades either side of 1800 saw a multifaceted change in how Europe's natural philosophers thought about the physical state of the earth. The older approach was predominantly Newtonian in spirit. Competing "theories of the earth" attempted to elucidate the natural laws whose operation must both explain the world's present-day form and determine the course of its future development. But during a period centered in the revolutionary decade of the 1790s, antiquarian history was imported into the study of the earth. By analogy with the archaeological exploration of the classical world, European savants came to search "nature's archives" for "nature's monuments" and "nature's inscriptions." What that implied was the simple but fertile idea that things could have turned out differently. The old supposed laws and principles might adequately label the forces acting on the world, but they could not explain how those processes had in fact played out over time. The course that affairs had taken should instead be regarded as a one-time-only matter of happenstance, impossible to predict from first principles. It was an epic adventure, a living drama (here lay the idea's "spiritual" quality) in which mighty forces had contended. The world as it now existed had been shaped by the outcomes of those clashes, which were as much a matter of fortune as the outcomes of battles between human armies. Landscapes and rock formations were so many monuments to those great events. This new way of seeing was, in short, "historical" rather than just "temporal." It remains a fundamental principle of the earth sciences today.

The Anthropocene, as I imagine it in this book, is a way of making current environmental change tangibly a part of this immense and circumstantial pageant. The Anthropocene does not put an end to natural history. On the contrary, it locates the present firmly within the geohistorical narrative first conceived of in the time of the French Revolution. Human history, having first been seen as an enlightening analogy for earth history, can subsequently be recognized as an increasingly forceful participant in it. Indeed, the idea of the Anthropocene is best thought of as part of a broader modern development in earth science, a development that has reinvigorated the historicizing impulse of the 1790s.

Some of the chancy, violent happenings of deep history were actually reconstructed in the age that first made conceptual room for them: before the middle of the nineteenth century Georges Cuvier convinced the scientific world of the controversial thesis of species extinction, and Louis Agassiz showed that the landscape had been pummeled into its present shape by a relatively recent ice age. But many other key elements of the drama were not. Darwin's theory of natural selection (as opposed to Jean-Baptiste de Lamarck's much less deeply historical "transformism") would not be made public until the second half of the century. More broadly, the new geohistorical sensibility depended upon the thesis that the earth was extremely ancient but not eternal; and under the influence of Charles Lyell— the thinker most frequently regarded as the father of geology—acceptance of this long timescale had become fused with the acceptance of a "gradualist" constraint on the earth's historicity. That is, belief in deep time had seemed to be of a piece with believing that the world had always been formed by slow physical processes, like sedimentation and erosion, that we can see continuing to operate in the present.

Only in recent decades has gradualism finally been pushed aside. Both slow, continuous processes and cataclysmic change are now typically credited with decisive roles in the geohistorical drama. That new paradigm is the outcome of a "neocatastrophist" turn in the late twentieth century.[9] It means that the historical impulse in earth science has become livelier than ever. No single breakthrough is responsible for this change of approach, but the outstanding instance of the neocatastrophist perspective is the recognition, which established itself over the course of the 1980s, that the land dinosaurs were abruptly wiped out (or at the very least given their coup de grâce) by the impact of a comet or asteroid, the Chicxulub bolide. A classic earlier case was the acceptance of another seemingly outrageous theory, that the northwestern United States had been shaped by apocalyptically vast floods at the end of the last ice age. Cognate developments can be found in many other fields: in paleontology, for instance, Stephen Jay Gould and Niles Eldredge's model of "punctuated equilibria" in evolution, whereby spells of relative stability are occasionally interrupted by rapid allopatric speciation; in paleoclimatology, recognition of the extreme swiftness of global warming at the end of ice ages, and of the astounding jumps between an ice-covered "snowball earth" and a ferocious greenhouse climate around the time of the origin of multicellular life; in paleoanthropology, the contentious theory of a human population bottleneck associated with the Toba supereruption; and so on.

The idea of the Anthropocene belongs in this context. Construed as a neocatastrophist concept, it is the reverse of the belief that industrial modernity is an alien power that has crash-landed in the still pond of the world. Instead, it makes the current suite of ecological changes the latest in an array of upheavals—some of them desperately harmful to the whole biosphere—that have emerged and reverberated within earth's systems. Framing the environmental predicament in that way ties it to the geohistorical narrative that has been increasingly well understood over the last two centuries. And this in turn makes possible a kind of understanding that might, one way or another, contribute toward well-judged actions in the face of the crisis.

CYCLES, GYRES, AND FEEDBACK LOOPS

On the other hand, if the planet's fundamental biological and chemical processes are always so dynamic, it is reasonable to ask whether current human influences on the environment really stand out among other types of changes to the earth. The signal of human activities must be remarkably strong if it is not to be lost in the noise of geological time. Admittedly, it is not hard to find indications of how far human influence has spread. The current depth record for a piece of litter is held by a can that was noticed by a submersible more than seven kilometers underwater in the Ryukyu Trench of the northwest Pacific. At the other end of the world, on the South Shetland Islands of Antarctica, the two native seed plant species have now been joined by a third, invasive one, annual bluegrass, a robust turfgrass elsewhere favored for putting greens.[10] But drink cans may decay, and the species compositions of ecosystems are always changing. Examples like these do not tell us what proportion of the planet, in total, has been shaped or ravaged by human activity. Perhaps, though much is taken, much abides: the highly visible scars might be far outweighed by the healthy tissue. After all, the earth is a big place. Before we can start to decide whether a new geological epoch might really be coming into being, or whether "the Anthropocene" is a plausible name for it if so, we need to assess the scale of contemporary anthropogenic changes to the earth system, measured against the size of the planet as a whole.

At 50 kilograms per person, the total mass of all living human beings is more than 350 million metric tons. When comparing quantities of biomass, however, scholars do not usually concentrate on raw figures like these (most of that 350 million tons is just water). Instead they calculate the mass of the carbon that organisms' bodies contain, because carbon is the key ingredient of all life on earth. Looked at this way, human bodies contain over 64 million tons of carbon, while, at the beginning of this century, domesticated animals contained more than 120 million tons of carbon, mostly in cattle. In contrast to those large numbers, the total for *all* wild terrestrial mammals, from armadillos to elephants, was about 5 million tons. The divergence is only increasing. On the other hand, all mammals combined are hugely outweighed by the land plants, which might hold a total of some 550 billion tons of carbon. There too, however, human influences have been transformative. Anthropogenic deforestation and land use change mean that that figure, large as it is, represents an estimated decline of 200 billion tons in two centuries.[11]

The situation at sea is more uncertain. The ocean holds more than 1.3 billion cubic kilometers of salt water (that is, more than 1.3 billion billion metric tons), all but 0.03 billion cubic kilometers of it black as night, and the sediment layer beneath—which is ten kilometers thick in places—constitutes another gigantic habitat. The microbial ecosystems of the dark ocean and its bed really have been largely beyond human perturbation so far. They are enormous, but they are not so big as to dwarf the human-dominated systems on land, and perhaps not so big as to match them. A high estimate puts the quantity of carbon contained in simple prokaryotic organisms (bacteria and the like) living in the sediment at the bottom of the oceans at 300 billion tons. A more recent estimate reduces that figure almost tenfold, however, while the number of prokaryotes living in the dark ocean itself might be not much more than a tenth again of that lower figure. The dark ocean also contains some fish and other macrofauna, of course, but the reality is that it is comparatively ecologically empty. Biomass concentrations decrease by 99 percent beneath the relatively narrow light-receiving upper layer, where two-thirds of the ocean's primary biological productivity takes place.[12]

The ocean's fertile regions are the surface waters and continental shelves, and human impacts there have been much more purely destructive than on land. A conservative estimate near the end of the twentieth century had it that bottom-trawling vessels—which pulverize the seabed by dragging a heavily weighted net over it—covered an area equivalent to half of the continental shelf each year.[13] In total, 90 percent of marine species live on the continental shelves, and over a third of all primary biological production there is now co-opted by fisheries. But that is still far too little to sustain the fishing industry. Wild fish populations are in what looks like terminal decline, which is why the total global catch has been decreasing steadily since the late 1980s despite ever fiercer industrial harvesting. The North Sea has been virtually emptied: the number of large fish there—anything bigger than four kilograms—has fallen by more than 97 percent. Populations of sea cows, sea turtles, oysters, large whales, and large marine predators in general have all fallen by 85 percent or more, worldwide. Caribbean monk seals were once so abundant "that all of the remaining fish on Caribbean coral reefs would be inadequate to sustain them"; they are now extinct.[14] Caribbean coral cover itself has fallen by nearly four-fifths.[15] Predatory fish biomass in the North Atlantic, already severely depleted by the middle of the twentieth century, decreased by another two-thirds by the century's end; whale biomass in the Southern Ocean fell sevenfold in the eight decades to 1985. The implication is that "marine capture is at or near its theoretical limit": most fisheries have now been overexploited, and perhaps soon more fish will be farmed (or caught to feed farmed fish) than can be hunted.[16]

Major Holocene marine ecosystems were centered on extensive food webs in which sharks, cod, tuna, and the like preyed on diverse populations of shoaling forage fish. The world ocean's emergent new state is characterized by a flattened food pyramid and by boom-and-bust population cycles of algae, dinoflagellates, and jellyfish. The destruction of shallow-water habitats and the rifling of deep oceans by subsidized fishing fleets; the massive summertime dead zones around the continents produced by the runoff of subsidized fertilizers and pesticides; the multiplication of ticks, parasites, and invasive species; ocean acidification; contamination from aquaculture; the suppression of upwelling deep water through warming of the upper ocean: the synergies between these effects favor opportunist scavenging species whose abundance is controlled by diseases and seasonal resource exhaustion, at the expense of long-lived predators and intricate reef communities.

In sum, the total number of nondomesticated vertebrates alive in the world (individuals, not species) declined by some 52 percent over the period 1970–2010. Half of all wildlife. Three-quarters, among freshwater animals.[17]

So much for the broad trends in population structures. What about the physical surface of the planet? Overall, a minimum estimate for the proportion of ice-free land currently devoted to human activity is 29 percent, a figure that counts only settlements, croplands, and woods cut down for pasture. A broader definition yields an estimate that 7.8 percent of all non-icebound land is "densely settled"; 16 percent consists of essentially cropland ecosystems in which at least a fifth of the land is actually under cultivation; 33.5 percent is more or less loosely managed rangeland, at least a fifth given over to pasturage; and 17.5 percent is seminatural land, "significantly transformed" but less than one-fifth covered by urban or agricultural ground. Added together, that leaves just a quarter of the continents—mainly the least biologically productive parts—for what could reasonably be called "wildlands." In the United States, the total area given over to turfgrass lawns alone is roughly equal to the size of Iowa.[18]

Estimates of the amount of matter deliberately moved by humans each year in mining and construction range from 30 to 57 billion metric tons. That sum is comparable to the amount of sediment carried into the oceans by all the world's rivers: somewhere in the range of 8 to 51 billion tons annually.[19] But humans cause much greater movements of sediment in a less direct fashion, through the erosion of agricultural soil. All in all, the "net anthropogenic denudation rate" of sediment and rock is estimated at ten times that due to all other factors combined.[20] Not surprisingly, this has practical consequences. Soil erosion forces the abandonment of some hundred thousand square kilometers of cropland—an area the size of South Korea—each year, meaning that in the last fifty years erosion has caused the loss of productivity in a third, perhaps more, of the world's cropland. To that must be added the losses following from the overgrazing of pastureland and from deforestation. Subsistence farmers on marginal lands are the worst affected, and they have the least access to the petroleum-based fertilizers that must be pumped in to generate adequate food on the remaining land. Soil degradation combines with water shortages and aquifer depletion, climate change, bureaucratic parasitism, agricultural surplus dumping, and civil strife to drive a constant flow of humanity into the toxic peri-urban slums of the global South, where populations rose from 760 million at the start of the millennium to 862 million by 2012.[21] Thus soil erosion and social violence are deeply interlinked, a connection made vivid in 2014 when the landslides that followed floods in the Balkans unearthed and detonated mines left over from the civil war.[22]

Food production and anthropogenic erosion now dominate some of the fundamental biogeochemical cycles of the earth system, like the cycles of nitrogen and phosphorus, both of which are essential to all forms of life. Modern agriculture is made possible by perhaps the largest modification of the planet's nitrogen system "since the major pathways of the modern cycle originated some 2.5 billion years ago." Humans have almost doubled the global rate of production of reactive nitrogen through their cultivation of legumes, fossil fuel burning, and, primarily, the manufacture of fertilizer through the Haber-Bosch process. Fertilizers also consume around 14 million metric tons of phosphorus each year, obtained by mining nonrenewable phosphate rock. Fertilizers make up the bulk of the 23 million tons of phosphorus added to cropland each year, while a larger quantity, 33 million tons annually, is removed, partly in food but mostly (20 million tons) through soil erosion. That erosion is the principal cause of eutrophication: blooms of waterborne phytoplankton gorge on the phosphorus and suck in the surrounding oxygen, turning affected waters luridly green and uninhabitable. Worldwide, two or three times as much phosphorus flows into the oceans as in preagricultural times. Thus, a one-way mining–erosion–eutrophication flow has broken open the essentially closed loops of the phosphorus cycle. The flow will exhaust known phosphate reserves in about 120 years at current speeds, and in the meantime the soils of countries where the fertilizers are unaffordable suffer from acute phosphorus deficiencies.[23]

Sewage from animal feedlots produces still thicker concentrations of phosphorus. The farm animals of the United States alone generate about 40 metric tons of shit per second.[24] Smithfield, America's largest pork producer, directs the outflow from its facilities into what it calls "lagoons," some more than a hectare in area and over nine meters deep. The shit is routinely sprayed on surrounding fields and intermittently lost in seepage, ponding, and immense river and wetland spills. It contains hydrogen sulfide, cyanide, various heavy metals, and "more than 100 microbial pathogens that can cause illness in humans," as well as stillborn piglets, afterbirths, and residues of the drugs that keep the tortured pigs alive. The complex chemical composition of industrial pig waste means that the stinking lagoons are not brown but pink. (Smithfield responded to an exposé of its practices by saying that "a pink lagoon is a healthy lagoon," and by boasting that "there are no known examples of the federal government forcing, or even asking, Smithfield to modify waste lagoon systems.")[25]

Far out in the ocean, one sampling of the surface waters of the North Pacific Central Gyre, better known as the great Pacific garbage patch, found that plankton were outweighed six times over by particles of plastic, which pass up the food chain to toxic effect through filter-feeding organisms, fish, and birds. Anthropogenic carbon dioxide emissions have increased the concentration of hydrogen ions in surface waters—that is, the water's acidity—by a menacing 26 percent.[26] Extensive research has been devoted to the possibility that the ongoing melting of the Greenland ice sheet could bring an abrupt halt to the downwelling of water in the North Atlantic, the hub of the entire ocean circulation system that drives the way in which energy moves around the surface of the earth.

Given the scale of these changes to the structure of the biosphere and the surface of the earth, it might be a surprise that fewer than 1 percent of all species are thought to have been altogether exterminated in recent times. Even though extinction casualties in the last few decades include two of the three Indonesian tiger subspecies, along with subspecies of ibex, otter, tortoise, and rhinoceros, and species of seal, sea lion, and dolphin, the present wave of extinctions is, so far, small compared to the five great extinctions in the history of complex life on earth, those when the number of living species fell by 75 percent or more.

But the headline extinction rate tells only a small part of the story. Many more species have undergone population crashes, or disappeared from large parts of their ranges, or otherwise become at risk of extinction. Suppose, then, that all the species currently recognized as "critically endangered," and no others, were to become extinct over the next hundred years—by no means a pessimistic scenario—and that extinctions then continued at the same rate. In that case, it would take roughly 1,500 years for three-quarters of all mammal species to be wiped out, 2,300 years for three-quarters of birds, and 890 years for three-quarters of amphibians. Humans are still some way from causing a sixth great obliteration of genetic diversity, in other words, but they are on course to do so within a geologically brief spell.[27] So far, counting both the critically endangered and those whose disappearance seems less imminent, the Red List prepared by the International Union for the Conservation of Nature identifies 25 percent of mammal species, or 1,143 in total, as threatened with extinction. The figure for birds is 13 percent (1,308 species), and for amphibians, 41 percent (1,950 species). In total, over 21,000 of the species whose extinction risk has been evaluated have been listed as under threat.[28] Gigantic and expanding pools of a few caged produce species have taken the place of those shrunken wild populations. The dominant tendency is the obliteration of diversified ecosystems and their replacement by close-packed monocultures, swarmed by pests and infections, through which maximal nutrient flows can be driven. Music becoming white noise.

Species extinction and human immiseration go hand in hand. Some two hundred species of fish that were unique to Lake Victoria have been eradicated since Nile perch were introduced there in 1954. The perch were intended to make possible a large commercial fishery, and they have done so. But around the lake's shores malnutrition is so severe that four-tenths of children experience stunted growth. The reason is that capitalizing on the giant perch requires expensive vessels, processing factories, and access to rich-world markets. The owners of the gear have absorbed the profits and proletarianized the fishermen. The children of the lake are cared for and given food mainly by their mothers, but women are largely denied access to the fishery and restricted to marginal activities like the processing of fish waste, or sex work in the brothels of the jerry-built shoreline boomtowns. In this way, the slow-burning ecological and humanitarian disaster of Lake Victoria entangles the feeding habits of Nile perch with the workings of ethnicity, nationality (the lake is shared between Tanzania, Uganda, and Kenya), class, and gender.[29]

And then there's climate change. Over the course of two decades of fatuous hardball in international climate change negotiations, colossal extra quantities of heat—equivalent, infamously, to four Hiroshima bombs each second—have amassed in the oceans and atmosphere. Most of Colombia's glacier cover has now gone, Mount Kilimanjaro has lost 80 percent of its snows, and the Chacaltaya glacier in Bolivia, once the world's highest ski run, has melted away completely. The 2479 square kilometers of the Larsen B ice shelf in West Antarctica, an area the size of Luxembourg or seven Grenadas, took less than three weeks to disintegrate into a swarm of icebergs when its warming reached a tipping point in early 2002.[30]

As yet, however, we have seen only the beginning of the climate system's jump from one state to another. Anthropogenic fossil-fuel burning and land use change released 555 billion metric tons of carbon into the atmosphere in the form of CO_2 between 1750 and 2011. Two hundred forty billion tons stayed in the atmosphere; half of the rest was reabsorbed by terrestrial ecosystems, and the other half was absorbed by the oceans.[31] For comparison, the end of the last ice age saw about 200 billion tons of carbon added to the atmosphere from the oceans. That was the linchpin—although not the sole cause: then as now, many positive feedback mechanisms were involved—of a change in the planet's temperature of at least 4.9°C between twenty-two thousand and eight thousand years ago. The older date represents the last glacial maximum, when ice sheets, more than three kilometers thick in central Canada, covered three-tenths of the world's land surface and Arctic tundra stretched from the north face of the Alps to Mongolia. The more recent date was the warmest point of the early Holocene, when the world was about 0.7°C hotter than it was just before the Industrial Revolution.[32] At present the world's total indicated fossil fuel reserves hold another 780 billion tons of carbon, and the global economy rests on the curious supposition that all of it will shortly be extracted and burned.[33]

It is painful to say that efforts to keep climate change to even minimally tolerable levels may well be futile by now, if only because that sounds like a self-fulfilling prophecy. But the feedback mechanisms already triggered mean that no human power whatsoever can halt the changes that are now under way. Across western North America the heat is exhausting pine trees and invigorating mountain pine beetles. A plague of the beetles an order of magnitude larger than any previous one has persisted since the end of the twentieth century. Hundreds of thousands of square kilometers of pine forest from Colorado north to the edge of the Yukon have been desolated, and as the forest dies it releases its stores of carbon, raising temperatures still further. This single beetle outbreak might add something more than a quarter of a billion metric tons of carbon to the air.[34] On its own, this is, by present standards, a drop in the bucket: one ugly example amid countless others.

The largest climate feedback machines, where tens of billions of metric tons of carbon are at stake, lurk in tropical rain forests and peatlands and in the far northern tundra. Gargantuan stocks of carbon have accumulated in permafrost regions, where millennia of cold have hampered the ordinary recycling of organic matter. But global warming is fastest at high latitudes, and the tremendous frozen peat bogs of Siberia are melting into sodden plains from which carbon is bubbling forth, often in the form of methane (a much more intense greenhouse agent than carbon dioxide). The cruel thing about these feedbacks is that their effects will be most pronounced if direct anthropogenic emissions are limited. Suppose that the earth proves to be markedly sensitive to climate forcings, and that, in response, fossil fuel emissions are after all slashed heroically in the next few years. In that case, self-perpetuating permafrost disintegration may on its own add 0.73°C of global warming by the end of this century, and 1.62°C within three centuries—this according to a conservative model that assumes all the emissions are only of CO_2, not methane. The climate system is bristling and seething with new sources of energy. High-speed global warming is not an imminent threat but the new condition of the earth. The first fourteen years of the twenty-first century were all among the sixteen hottest since instrumental records began in 1880. The two interlopers were the El Niño years of 1997 and 1998.[35]

Some of the transformations now taking place are, precisely, biblical in scale. Over a quarter of Hong Kong's urban land, sixty-seven square kilometers of it, has been reclaimed from the sea. China's South-North Water Transfer Project will, if completed, carry forty-five billion cubic meters of water a year across a vast stretch of the country, becoming the single largest construction project the world has ever seen.[36] The first hurricane ever recorded above the warming waters of the South Atlantic made landfall in Brazil in 2004. Above the most inaccessible land on earth, East Antarctica, snowfall is increasing in the context of a general poleward shift in precipitation patterns.

And God said: *who* shut up the sea with doors, when it brake forth, *as if* it had issued out of the womb? Who hath divided a watercourse for the overflowing of waters, or a way for the lightning of thunder; To cause it to rain on the earth, *where* no man *is; on* the wilderness, wherein *there is* no man?

And Man said: I did, actually.

CHAPTER TWO

Versions of the Anthropocene

Since the beginning of this century, one way of referring to the crisis that I described in the previous chapter has become ever more popular and ever more controversial. The word *Anthropocene* has come into fashion, and in doing so it has picked up a variety of incompatible meanings, each implying different concepts and commitments. The word's complexity means that there is little to be gained by talking about "the Anthropocene" without specifying which version of it you mean. It is especially unfruitful to denounce the word in blanket terms if your real target is only one particular way of using it. Even so, and for understandable reasons, the concept of the Anthropocene has recently been indicted wholesale by a number of writers. Hostile critics have accused it of a domineering universalism: of downplaying the differences between Albertan oil barons and Malagasy subsistence fishers by suggesting that it is human beings in general who are responsible for ecological degradation. Thinking historically about how planetary systems operate, however, sheds a different light on the central issues in that controversy. I believe that one version of the Anthropocene in particular might prove to be a useful and enabling one for contemporary green politics. The stratigraphic approach to the Anthropocene, which contemplates

introducing the word as the name of a new interval in the geological timescale, provides a way of thinking about power relations as they exist both
among human beings and between all kinds of geophysical forces.

"WE'RE NOT IN THE HOLOCENE ANYMORE"

As witnesses tell the story, it goes like this. At a conference on earth system
science outside Mexico City, early in the last year of the twentieth century,
participants talked about the Holocene, the geological time span that offi-
cially includes the present day. One listener apparently felt a sudden, curious
revulsion. Paul Crutzen was an illustrious scholar of the earth sciences, most
celebrated for work that enabled the discovery of the hole in the ozone layer.
He had received the Nobel Prize in chemistry a few years earlier. Now he was
struck simultaneously by a novel idea and by a word to express it. "Stop
using the word *Holocene*," he told the conference delegates. "We're not in the
Holocene anymore. We're in the . . . the . . . the Anthropocene!" (The precise form of his words varies slightly from telling to telling.) The room fell
quiet; "everyone was shocked." Then a buzz of conversation arose. "Some-
one came up to Crutzen and suggested that he patent the term."[1]

The anecdote about Crutzen's impulsive declaration offers a seductively
memorable starting point for the study of the Anthropocene, a name evidently intended to mean something like "the human epoch." But historians
of science are constitutionally skeptical of Eureka moments, and the most
convincing accounts of how scientific paradigms change usually give low
priority to singular flashes of inspiration such as this. In this case, Crutzen
himself has worked to make it clear that the emergence of the concept was
more drawn-out and more complicated than the story about the conference
might suggest. He swiftly wrote up his vision of the Anthropocene in the
modest setting of the in-house newsletter of the International Geosphere-
Biosphere Programme (the organizers of the Mexico conference) for May
2000. That brief article was coauthored with Eugene Stoermer, an American

ecologist—a student, principally, of photosynthesizing algae—because Crutzen had learned that Stoermer had been using the same term informally since the 1980s. Together, Crutzen and Stoermer listed earlier analogues to their theory: George Perkins Marsh's prescient conservationist treatise *Man and Nature; or, Physical Geography as Modified by Human Action;* the declaration of the "Anthropozoic era" by the Italian geologist Antonio Stoppani in 1873; and the work of Vladimir Vernadsky, the pioneering theorist of the biosphere and of its culmination in a noosphere, or "sphere of reason." Later they adduced the biologist E. O. Wilson, and the science writer Andrew Revkin, who offhandedly proposed an "Anthrocene" age in the early 1990s. Others have traced similar ideas back to the dawn of geological science, in the work of the eighteenth-century naturalist the Comte de Buffon.[2]

Crutzen and Stoermer's article in the *IGBP Newsletter* made clear the grand scope of their idea. Barring some global catastrophe, they wrote, "mankind will remain a major geological force for many millennia, maybe millions of years to come." Thus, "it seems to us more than appropriate to emphasize the central role of mankind in geology and ecology by proposing to use the term 'anthropocene' for the current geological epoch." In their view the new epoch began in the late eighteenth century, when an appreciable rise in atmospheric methane and carbon dioxide levels began the season in which "the global effects of human activities have become clearly noticeable." They added, "Such a starting date also coincides with James Watt's invention of the steam engine in 1784."[3] It would turn out that the question of how to define the beginning of the Anthropocene could not, by any means, be resolved as easily as that. Even so, this first sketch of the human epoch is a document of enduring significance.

In January 2002 Crutzen recapitulated his view of the Anthropocene in a far more widely circulated journal, *Nature*. It is this article that best marks the emergence of the concept into widespread scientific awareness. To date, it has been cited well over a thousand times. Although it was even shorter than the *Newsletter* discussion and said many of the same things, Crutzen found room for two important new points. The first was a brief acknowledgment, missing from the first piece, that the changes apparently bringing about a new epoch "have largely been caused by only 25% of the world population." The second was a bold gesture toward the kind of "environmentally sustainable management" that might be suited to the Anthropocene: "This . . . may well involve internationally accepted, large-scale geo-engineering projects, for instance to 'optimize' climate."[4] Both of these themes—the need to recognize that people in different parts of the world have made very dissimilar contributions to global change, and a distinct inclination toward geoengineering as a way of dealing with global warming—would remain prominent in discussions of the Anthropocene.

Crutzen's seminal *Nature* article is the canonical statement of the first version of the Anthropocene. The tone is clear, humane, and confident; cognizant of the power of technology; socially engaged, although not polemical; pessimistic but not despairing in its assessment of the state of the planet; and magisterial in the way that it evaluates the sum of human environmental influence. The concept of humanity's epoch struck a chord, and the idea quickly began to circulate, filtering into a whole range of earth science disciplines and, before long, beyond them. Crutzen's term began to appear in articles about human geography and geopolitics and in books for general audiences by environmental writers.

Many readers have continued to find things of value in the idea of the Anthropocene as it stood in those two early articles. But it is essential to any serious engagement with the Anthropocene to recognize that Stoermer's and Crutzen's first brief sketches do not by any means represent the only possible version of the concept, its unchanging real essence, or its true scientific meaning. On the contrary, the idea has been fissiparous from the start. Different fields have received it in various different ways. We can perhaps speak of more or less mainstream accounts—provided we keep in mind that the consensus about what constitutes the mainstream can alter rapidly—but no single version of the Anthropocene can reasonably be described even as a generally dominant one.

The clearest illustration of that principle is the fact that Crutzen's own ideas have changed significantly. In collaboration with the environmental historian John McNeill, among others, he came to argue that the Anthropocene began in a more piecemeal fashion than he had proposed at first. In his revised account, atmospheric CO_2 continues to serve as "a single, simple indicator to track the progression of the Anthropocene," but the new epoch is said to have emerged in two stages. Stage 1 began in "the 1800–1850 period," with the breakthrough development of fossil-fueled industrialization in Britain. But the revised account acknowledges that, until the middle of the nineteenth century, CO_2 concentrations did not in fact pass outside the range within which they had been fluctuating for ten thousand years. More generally, the new model characterizes stage 1 human environmental impacts as burgeoning rapidly rather than growing explosively. Truly vertiginous economic growth, in this account, was incipient in the period of high Victorian liberalism from the 1850s onward but was held back by the world wars and Great Depression. Stage 2 of the Anthropocene, then, begins with a "Great Acceleration" after 1945, when the momentum of the "human enterprise" multiplied precipitately. This analysis is based on a dozen much-reproduced graphs that show levels of population, worldwide GDP, fertilizer consumption, paper consumption, foreign direct investment, international tourism, and so on undergoing a nearly vertical takeoff in the middle of the twentieth century. For Crutzen and his colleagues, these graphs represent the Anthropocene's transition from its larval to its adult stage, because they correlate with the exponential increase of human pressures on "Earth's life-support system."[5] The Anthropocene in its full sense is even younger than it had at first appeared.

NEW MEANINGS

Crutzen's two versions of the Anthropocene were not alone. In the decade after the term was coined it was put to many other uses, and these embraced a far wider range of definitions. While Crutzen was moving the Anthropocene's start date closer to the present, another distinguished student of the planet's atmosphere was carrying it much further back. "The Anthropocene actually began thousands of years ago," William Ruddiman argued, "as a result of the discovery of agriculture and subsequent technological innovations in the practice of farming." Ice-core records seem to show anomalous greenhouse gas concentrations during the current interglacial, or warm spell between ice ages, compared to preceding ones. Guided ultimately by changes in the earth's position relative to the sun, these concentrations should have reached a peak not long after the last ice age and then trended downward, but instead CO_2 levels (from eight thousand years ago) and methane levels (from five thousand years ago) show a small, unexpected rise.

Humans, in Ruddiman's view, were responsible. Forest clearance in Eurasia for agriculture and fuel explains the CO_2 anomaly, an extra forty parts per million in the preindustrial atmosphere, and East Asian rice paddies produced the additional methane. Thus, preindustrial farmers unwittingly postponed the next ice age, which would otherwise have begun to take hold in northeast Canada thousands of years ago.[6] It is a startling hypothesis, and one that has provoked much debate. At present the developing consensus is against it. Interglacials other than those on which Ruddiman focused provide better analogues to the present one, and they seem to make the changing composition of its air appear much less anomalous. Ruddiman himself, however, continues to hold the line on his early Anthropocene hypothesis. This version of the Anthropocene does have one important similarity to both of those proposed by Crutzen. In all three cases, the new epoch is understood as the time since human activities took atmospheric carbon dioxide levels outside the range they would have occupied in humans' absence. The enormous difference in the dating of that change, however, points to the two scholars' radically opposed assessments of how humans have influenced the way the world works.

Many other early Anthropocenes, distinct from Ruddiman's, have also been proposed. One extreme in the dating of the epoch, at least so far, places its origin at 1.8 million years ago, at an earlyish date for the mastery of fire by hominins. The rationale is that this was the crucial technological achievement—because cooking renders the digestion of animal protein more efficient—that allowed for the evolution of a new line of large-brained, tool-using apes. Elsewhere the Anthropocene has been defined as the interval since the extinction of most genera of megafauna over most of the world, between about fifty thousand and ten thousand years ago, at the hands of newly arrived human hunters. Another, less bleak option identifies its onset with the domestication of animals and plants, making the Anthropocene approximately coeval with the Holocene and just a little older than Ruddiman suggested. Or the Anthropocene could be two millennia old and recognizable in the changes to much of the world's soil—through manuring, irrigation, terracing, and so on—associated with the empires of the time of Christ.[7]

Decisions about historical periodization very often encode deep interpretive commitments. In this case, the general rule is that the earlier the proposed starting date for the Anthropocene, the more emphasis its proponents place on human actions themselves, as opposed to the ecological consequences that follow from them. Bruce Smith and Melinda Zeder—supporters of an Anthropocene defined by species domestication—are the thinkers who take this position most explicitly. Smith and Zeder believe that one should identify the Anthropocene with the emergence of "significant human modification of the earth's ecosystems on a global scale," rather than looking only for "massive and rapid . . . human impact" like that seen in the past two centuries. This lower bar implies a conceptual reversal. In a word, "the focus should be on cause rather than effect, on human behaviour [rather than] environmental degradation." For them, the Anthropocene is a way of naming the whole interval during which humans around the world have significantly shaped or engineered their habitats. Whereas Crutzen and Stoermer proposed the term as a framework for assessing the general state of the planetary system, Smith and Zeder employ it as a heuristic device for "gaining a greater understanding of the . . . role played by human societies in altering the earth's biosphere."[8] Their focus is on the human capacity to change the world, not on the changes themselves.

Another version of the Anthropocene came about as the concept was picked up in the humanities. For scholars of politics and culture, the most obvious questions to ask are less about the origins of human environmental impact and more about the implications of the Anthropocene for social organization. The landmark contribution in this vein has come from the historian and postcolonial theorist Dipesh Chakrabarty. For Chakrabarty, the Anthropocene's significance lies in the fact that postcolonial and Marxist scholars' radical critiques of globalization, capitalism, and imperialism are all inadequate in confronting the idea of a new geological epoch. No matter how compelling they are on their own terms,

> these critiques do not give us an adequate hold on human history once we accept that the crisis of climate change is here with us and may exist as part of this planet for much longer than capitalism. . . . A critique that is only a critique of capital is not sufficient for addressing questions relating to human history once the crisis of climate change has been acknowledged. . . . Whatever our socioeconomic and technological choices, whatever the rights we wish to celebrate as our freedom, we cannot afford to destabilize conditions (such as the temperature zone in which the planet exists) that work like boundary parameters of human existence.[9]

The Anthropocene, in this reading, means recognizing the fact that the environmental crisis constitutes a major challenge for the kind of politics that resists the inequities of the existing world order. Yet Chakrabarty is not so cynical as to say that analyses of social and economic injustice must be abandoned under the pressure of the Anthropocene. "Critiques of capitalist globalization have *not*, in any way, become obsolete in the age of climate change," he insists. But climate change means that on their own they are no longer enough. What he proposes instead is a double perspective, an attempt "to mix together the immiscible chronologies of capital and species history." Historians still need to tell "the story of capital, the contingent history of our falling into the Anthropocene," with its themes of liberation and injustice and its chronological range of several hundred years. At the same time, however, they now need to trace another longer, deeper history of humankind as a species, and of human interactions with the rest of the planet's life, over timescales of thousands and millions of years.

In the course of making this argument, Chakrabarty turned back to the first presentation of the idea of the Anthropocene. He cited, with qualified approval, the last two sentences of the article in the *IGBP Newsletter* with which Crutzen and Stoermer introduced the concept. That article concluded with a flourish. "To develop a world-wide accepted strategy leading to sustainability of ecosystems against human induced stresses will be one of the great future tasks of mankind," Crutzen and Stoermer had written. "An exciting, but also difficult and daunting task lies ahead of the global research and engineering community to guide mankind towards global, sustainable, environmental management."[10]

When Chakrabarty reproduced those sentences, the idea of the Anthropocene ran into trouble—because there is much to take issue with in the political standpoint implied by Crutzen and Stoermer's words. A truly global community of researchers and engineers can hardly be said to exist, given how unevenly the money to support scientific research is distributed across the world. And it is plain that no one appointed such a community to the task of guiding "mankind" anywhere. The relevant researchers are a disputatious body of thinkers and investigators, not vatic universal steersmen. Crutzen and Stoermer's hope that a single strategy for sustainability will be accepted worldwide likewise appears utopian at best. Who would have the power to declare that the world had accepted any given strategy, and what would happen to those who remained unwilling to agree? Perhaps most importantly, the pair's proposals for "environmental management" seem like exactly the way of thinking that the Anthropocene undermines. How can we plan for the "sustainability of ecosystems *against* human induced stresses" once we have recognized that most ecosystems have already been profoundly remodeled, with human activities placed in a central role? "Human induced stresses" are a part of the system, like the stresses brought on by the changing of the seasons. The managerialist belief that it is humankind's duty to regulate the natural world from the outside sits oddly with the recognition that the fundamental biogeochemical matrices of the planet are now fused with human activity. But when Chakrabarty repeated Crutzen and Stoermer's words in an essay on the future of postcolonial studies, it became clear that this line of thought was a tenacious feature of the discourse on the Anthropocene.

Chakrabarty's brief for a dual approach to environmental analysis, linking a critique of capitalist globalization to a longer history of humans as a species, was plainly an exciting one. But several of his readers came to suspect that the two strands of his approach could not readily be woven together in the way that he envisaged. Intertwining them sounds welcome, but what if the latter (species-based) one just encircles and subsumes the former (political) one? Chakrabarty's initially unwary invocations of the "shared catastrophe that we have all fallen into" as the basis for a "new universal history of humans" suggested that that danger was real. "Unlike in the crises of capitalism," he wrote, "there are no lifeboats here for the rich and the privileged (witness the drought in Australia or recent fires in the wealthy neighborhoods of California)."[11] A rich Californian whose house burns down faces emotional trauma and a home insurance excess, but she hardly shares the experience of a drought-stricken cattle pastoralist in South Sudan. "No lifeboats"? As Chakrabarty's critics have been happy to point out, this is untrue both literally and metaphorically. The militarization of disaster areas like Katrina-struck New Orleans, and the financialization of catastrophe through disaster reinsurance, have already proved capable of preserving—indeed reinforcing—capitalist hierarchies in zones of ecological emergency. In the eyes of his critics, then, Chakrabarty had been drawn away by Crutzen and Stoermer's seductive idea from some of the indispensable tenets of postcolonial studies.

THE BACKLASH

Chakrabarty's groundbreaking discussions of the Anthropocene have become a lightning rod for attacks on the whole concept of the human epoch. In the years since his first essay on the subject appeared in 2009, the idea of the Anthropocene has become both much more widely employed and much more widely criticized. The concept started to reach the mainstream in 2011. A collection of essays on the topic appeared that year in a themed issue of the *Philosophical Transactions of the Royal Society*. There, Crutzen and his collaborators reflected on how the word had spread over the previous decade. "Since its introduction," they wrote, "the term Anthropocene has become widely accepted in the global change research community, and is now occasionally mentioned in articles in popular media on climate change or other global environmental issues."[12] Had they been writing a few years later, they could have noted much more than such occasional mentions. The year 2011 itself saw a flurry of major conferences, as well as enthusiastic feature articles in *Science*, *National Geographic*, the *Economist*, and elsewhere, and the term began to crop up regularly in newspapers for the first time. Since then, there have been museum exhibitions and radio programs, academic research networks and chapters in textbooks, and, most remarkably, no fewer than three new scholarly journals: students of the new epoch may now turn to the *Anthropocene Review* (which has swiftly become the leading forum for discussion of the concept), *Anthropocene*, and *Elementa: Science of the Anthropocene*.

The Anthropocene has become fashionable in academic circles—very fashionable, in fact. In principle any discussion of anything that has taken place in the last few hundred years or so can be tagged with the phrase ". . . in the Anthropocene" and thereby made to sound (however transiently) up to date. The organizers of the latest conferences on the topic struggle to accommodate presentations of the most diverse and miscellaneous kind. Various words have been coined on the Anthropocene model: *Capitalocene*, *Sustainocene*, *Cosmoscene*, *Econocene*, *Homogenocene*. Some scholars, inevitably, have even decided that the time has come to speak of the "post-Anthropocene." And as the word itself has come into prominence, so the backlash has developed. If the concept has been associated with the idea that the whole world must be "guided" into adopting a single approach to environmental management, or with the claim that global warming's floodwaters will bear "no lifeboats" for the rich, then some suspicion is understandable. The opponents of the Anthropocene have warned that the whole notion may be politically naive if not implicitly unjust, and may diminish rather than improve the chances of equitable and efficacious responses to ecological crisis.

The argument against the Anthropocene is by now well established.[13] In two words, the accusation is that the idea is universalist and technocratic. It is universalist because it makes it sound as if we are all in this predicament together. It neglects humanity's division into a multitude of unequal social groups, and the ways in which wealth, nationality, ethnicity, gender, class, and so on mediate the relationships between those groups. In its simplifying view, the human species—the *anthropos* in general—becomes instead an abstract, homogeneous mass, collectively damaging the planet through vaguely defined habits of industrialization, resource exploitation, and over-consumption. Those habits are supposed to put at risk the well-being of the whole human race, meaning that the only solution is to set aside class resentment and work together as one for the greater good of the whole. The Anthropocene wrongly implies that humanity is united in culpability, in vulnerability, and in the need for self-protection. For the opponents of the concept, that makes it an essentially bourgeois idea. It performs the arche-typal bourgeois maneuver of representing the sectional interests of a single group as being in everybody's interests. It comforts the prosperous with the thought that blame attaches collectively to all human beings. Thus it lends itself to a blinkered preoccupation with overpopulation as the supposed root of all the world's ills, which means blaming the poor for a crisis to which they have in fact contributed very little.

According to this critical view, scholars of the Anthropocene rely upon a simplified, one-step model of historical change whereby the Holocene epoch was replaced everywhere and all at once by a human-controlled earth sys-tem. Their attempts to attach a date to that transition are bound to prove futile, because in reality different parts of the world have undergone very different experiences of modernization and development at very different times. But the Anthropocene deals only in aggregate environmental conse-quences, pushing the subtleties of causation into the shadows. That makes it deterministic: it presents human nature itself, the technological impulse of the *anthropos*, as a full and adequate explanation for the course of history. The Anthropocene theory of history is correspondingly depoliticized and preoccupied with scientific inventions. It fetishizes the Industrial Revolu-tion as the sole origin of modernity; and in doing so, it misrepresents that revolution as simply a technological leap forward, neglecting industrializa-tion's economic underpinnings. In a skeptical analysis, this habitual occlu-sion by Anthropocene enthusiasts of the politics of empire and capitalism is itself a deeply political act. Choosing this way of understanding the crisis predetermines the kind of solutions that will be proposed: modernist, high-tech, top-down ones. It is no coincidence that Crutzen himself has been among the world's most prominent advocates for geoengineering. The discourse of the Anthropocene is technocratic because it makes it sound as if there is no alternative to the rule of experts. It is a counsel of despair, sacrificing freedom and wilderness to managerial diktat. It implies that if we are to survive, then those who make political decisions must defer to a scientific and technical elite, who can specify the objective physical constraints on how humanity may make use of its life-support systems.

For those reasons, its critics charge, the concept of the Anthropocene can be suspected of a general affinity with approaches to rationing—of carbon emissions, fish quotas, and so on—that suit the developed world much better than the poor. It is an intellectual bedfellow of those hapless regulatory regimes that seek to conserve ecosystems by commoditizing them, like ecosystem-services markets and the European Union's chaotic emissions-trading scheme. At worst, it could lend an air of respectability to a process of environmental triage that would calmly sacrifice Tuvalu and the Marshall Islands to the greater good. Most fundamentally, though, it plays a philosophical double game. It pretends to describe human beings merging into nature as a geologic force, but in fact it is deeply dualistic. In portraying humans as a unified species—as the bearers of a singular human essence—it singles them out from the rest of the world. It sets the Anthropocene, the artificial age made by humans, squarely against the entire natural history of the world that preceded it.

That, or something like it, has been the standard critique of the Anthropocene so far. Gerda Roelvink sums it up: "In their announcement of the Anthropocene, scientists are calling us to consider ourselves not as a number of different groups but as a single, universal, and transhistorical collective—as a species. . . . This understanding of species fits all too easily with the modernist assumption of human mastery over nature."[14] Humanity reduced to an undifferentiated species, and such species thinking as a warrant for depoliticized, technology-driven management of "nature": that is the danger of the Anthropocene.

Yet Roelvink's complaint can be turned straight back against her. Her accusation itself takes "scientists" for nothing other than "a single, universal, and transhistorical collective" serving the interests of human mastery over nature. As we have seen, the reality is that researchers who have worked with the idea of the Anthropocene are, emphatically, members of "a number of different groups" much given to disagreeing with one another and to changing their minds about things. Not all takes on the Anthropocene are the same. Some are universalist and technocratic. Others, I argue, are quite the contrary. The crucial flaw in the case against the Anthropocene has been a widespread failure to recognize that the word *Anthropocene* does not express any single, agreed-upon idea. Instead, by the time the backlash started the term had already taken on a considerable number of mutually irreconcilable senses. Worse still, critiques of the politics of the Anthropocene have mostly aimed their fire at the simplest and most sketchily formed version of the concept, the first-draft Anthropocene of Crutzen and Stoermer's original *Newsletter* article and Crutzen's brief 2002 follow-up in *Nature*. Otherwise thoughtful dissents have taken those two texts as adequate representatives of Anthropocene thinking as a whole, or even been so naive as to assume that what is found there *is* the theory of the Anthropocene, its only possible form. The stronger charges may well be unfair even to those texts, which are provisional and suggestive enough to be interpreted in several different lights. But in any case there is little to suggest that every possible version of the Anthropocene is politically compromised in the same way.

The skeptical responses to the Anthropocene from some theorists of politics and social difference need not persuade anyone to abandon the term, but they should make researchers who draw on it refine and sharpen their analysis. The charges that the idea of the new epoch might stigmatize the poor, pander to elitist technocratic fantasies, disguise political or historical realities, or work against equitable responses to environmental problems are unsettling ones. They demand to be taken very seriously indeed. Any worthwhile version of the Anthropocene has to be underpinned by a historically nuanced account of how power relations operate, both across the earth system as a whole and between human beings. Scholars working in fields like postcolonial studies can make a vital contribution to that way of framing the new epoch. In what follows I have kept in mind the radical critique of the Anthropocene and sought to avoid speaking of humankind as an undifferentiated whole.

After all, there is still plenty to gain from rethinking the meaning of the Anthropocene. Even if the term has already passed further into popular awareness than any other geological concept since plate tectonics, its rise is probably far from complete as yet. Certainly, it is a sufficiently large and bold idea to bear a level of scrutiny many times greater than it has received so far. At the time of this writing, journalists still never use the word for a general audience without glossing its meaning, and it was canonized by an entry in the *Oxford English Dictionary* only in the summer of 2014. The environmental crisis has pushed into the limelight a whole series of words and phrases that condense the meaning of various contested concepts. All problematic in their way, they nevertheless show how language has been reshaped in the effort to come to terms with that crisis: *global warming, acid rain, the hole in the ozone layer, biodiversity, sustainable development, carbon footprint*. As yet, all of those terms are far more familiar than *the Anthropocene epoch*, even though the latter is as far-reaching a concept as any. But it remains to be seen whether the Anthropocene will exert a galvanizing force on public debate, as talk of the ozone hole did in the 1980s, or if it will instead become distracting and even misleading, as has become the case with invocations of "sustainable development."

GEOLOGICAL LIFE

Worthwhile accounts of the Anthropocene will be ones that illuminate instead of obfuscating the patterns of human-caused environmental change. Those patterns are political through and through. To find a way forward, then, it might be best to go back to the most important texts in the controversy about the politics of the Anthropocene and to reread Dipesh Chakrabarty's work. No objection to the Anthropocene has yet found a way of dealing with the central challenge that Chakrabarty poses. He recognizes that resistance to current forms of capitalist globalization, and to their profit-driven exploitation of disempowered communities and vulnerable ecosystems, is a prerequisite for the creation of a livable and equitable world. But he insists—convincingly—that anticapitalist resistance is not sufficient to that end. The environmental disaster is bigger than capitalism. It destabilizes "boundary parameters of existence" that are independent of the logic of capitalism. It involves climate changes that will almost certainly continue for far longer than capitalism as we know it. It has been caused by industrializing socialist states as well as by the Western powers.

For Chakrabarty, the Anthropocene signifies the division and incompatibility between the two problems of "globalization and global warming." The latter problem exists on a deeper level than the former, and it affects humankind en masse. Just revealing potential ill effects of the Anthropocene narrative, as Roelvink and others try to do, does not get rid of that division. In other words, even if the Anthropocene really does have unwelcome political consequences like stigmatizing the poor and promoting antidemocratic techno-fixes, it still might accurately describe the grim implications of the fact that all living conditions on the planet are under threat. So it looks as if we are faced with an impasse. Justified hostility to the claim that "we're all in it together" versus justified recognition that equal fossil-fueled prosperity for everybody appears ecologically impossible. The time of modernity versus the epoch of humankind. The political history of capital versus the geophysical history of carbon. Is this deadlock the end of the road for the idea of the Anthropocene?

No. The alternative is to reconceive the Anthropocene not as a signal of built-in contradiction and conflict between radical social critique and species thinking but as the very concept that unites the two. The birth of the new epoch is, precisely, an opportunity to think about human and nonhuman power relations simultaneously. The way to revise Chakrabarty's analysis, and to take it forward in a new direction, is to question or destabilize a distinction that was implicit throughout his early work on the Anthropocene. This is the distinction between life and nonliving matter. That distinction seems at first to constitute a stable and unambiguous binary opposition; but in another light, life and nonlife appear only as different moments within the interwoven cycles through which the earth system functions.

We can see the importance of the life/nonlife distinction to Chakrabarty's thought in passages like the following: "Climate scientists posit that the human being has become something much larger than the simple biological agent that he or she always has been. Humans now wield a geological force. . . . Humans are biological agents, both collectively and as individuals. They have always been so. . . . But we can become geological agents only historically and collectively."[15] This conception lies at the heart of the story that Chakrabarty tells. The Anthropocene, for him, is the result of this rupture in which humans were transformed from merely "biological" agents into a power that is both biological and "geological." Marxist, postcolonialist, and environmental-justice thinkers deal adequately with humans insofar as they are biological creatures like all other species, but their approaches are inadequate to humans considered in their modern collective or "universal" form as a geological force. To put it more explicitly still: as well as our biological existence, "we now also have a mode of existence in which we—collectively and as a geophysical force and in ways we cannot experience ourselves—are 'indifferent' or 'neutral' . . . to questions of intrahuman justice." Humans have become a "nonhuman, nonliving agency," in "a collective mode of existence that is justice-blind."[16] The opposition that Chakrabarty set up between the Anthropocene and the politics of liberation stemmed fundamentally from the way he envisaged this historical switch from the biological to the geological, from the living to the nonliving.

This seemingly clear binary contrast is, nonetheless, an incomplete, temporary, and conditional one. We can see that by thinking about the agencies that actually participate in the earth system. It might appear that the planet's workings are (or that they were, before humans) made up firstly of an inanimate geological base or framework, composed of phenomena like plate tectonics, volcanism, climate, erosion, and sedimentation; and secondly of a decorative organic superstructure, both supported and determined by the geophysical realities to which it is obliged to accommodate itself. It is true, after all, that the earth's biomass is tiny compared to the mass of its atmosphere, water, or rock layers. Nonetheless, that way of thinking has to a great extent been swept away by recent students of the earth's systems. Earth, in contrast to dead planets like Venus, has remained far out of chemical equilibrium for billions of years. It does so because life has now been an integral part of the planet's makeup for more than three-quarters of its existence.

Living organisms are concentrated at the active interfaces between the atmosphere, hydrosphere, and lithosphere. And chemical processing in the biosphere is usually far more rapid than elsewhere, as organisms photosynthesize, eat, respire, excrete, and die. For those reasons, many of the main cycles through which the planet functions are *biogeochemical* ones in which life and inorganic processes are inextricably combined. The earth operates in entangled loops of carbon, nitrogen, calcium, oxygen, water, phosphorus, and so on. Those loops involve journeys that pass through living bodies or that are facilitated by organic processes. The erosion of rock is the deep, fundamental driver of the carbon cycle and is a basic part of the plate tectonic cycle. It is forcefully accelerated by the bacteria, lichens, and fungi that eat away at stone surfaces. The atmosphere contains reactive oxygen only because oxygen has been excreted by bacteria for more than two billion years. The production of free oxygen, alongside other biological processes, brought perhaps as many as two-thirds of the earth's (nonanthropogenic) minerals into being.

Living things shape rivers and coastlines by colonizing and stabilizing sediments. They accumulate into landscape-size geological features: soil, peat bogs, coal seams, limestone cliffs. The hydrological cycle involves plant transpiration, water capture in vegetation-dependent soils, and gas emissions from algae that inflect cloud formation. Ice ages seem to be brought about partly through the operation of a "biological pump." In this mechanism, small changes in the earth's position relative to the sun increase the heat differential between tropics and poles, so that stronger winds blow between them and carry more iron- and nutrient-bearing dust into the oceans. That dust fertilizes microorganisms whose calcium- and carbon-rich bodies and shells sink when they die (or when they are eaten and excreted), thus sequestering carbon from the atmosphere and chilling the whole planet. The albedo, or reflectivity, of a land surface depends upon the vegetation by which it is covered; its albedo partly governs temperature and precipitation levels, and these climatic factors in turn influence the evolution of the vegetation. The presence or absence of large herbivores can dramatically alter the ground cover. Thus those herbivores too are geological forces, just like earthworms and beavers.

In short, life has *always* been a geophysical force; equally, the geology of the earth, unlike that of Venus, has been influenced by the laws of biological evolution for an inordinate length of time. The "biological agents" to which Chakrabarty referred have always been "geological agents" as well, and it is a rare "nonliving agency" that does not have a trace of life about it. (Indeed, the very existence of life demonstrates that self-replicating systems can emerge out of inorganic chemical processes.) Living things on the one hand, and geophysical things like rocks and climate on the other, are, at root, inseparable parts of the ecological cycles that operate on and around the surface of the earth. Biological and geological phenomena are not two different kinds of being upon which two different regimes of politics might be founded. Although the birth of the Anthropocene does change the way in which the forces of life and of geophysics are arranged, it does not affect their underlying unity.[17]

Chakrabarty conceived of the human species as leaping across a divide from the biological to the biological-and-geological, and he proposed that one side meant politics whereas the other side meant both politics and apolitical collective action. A consideration of the makeup of the earth's biogeochemical systems obliterates that divide. What looked at first like a difference of kind between life and nonlife becomes only a difference of scale between kindred geophysical forces—and indeed Chakrabarty's own recent work has turned to focus more explicitly on such questions of scale.[18] The consequence is that the deadlock between politics and the Anthropocene no longer stands. As that deadlock vanishes in the stronger light of history, it becomes possible to see plainly both the drawback of Chakrabarty's analysis and the great importance of his central insight. He was right—and boldly pioneering—to declare that emancipatory politics in the twenty-first century must undergo a challenging alteration as the result of an upheaval in the geological condition of the earth. But it is not the case, thankfully, that such geological upheavals belong on a plane entirely different from that of the normal struggles for advantage that go on constantly between living things. On the contrary, the two have always been knitted together throughout the planet's ecological systems. And struggles for advantage between living things are what politics deals with.

Politics is the right mode in which to address geological problems after all. It need not be circumscribed or replaced by a geological way of seeing that treats species as undifferentiated wholes, because the geological way of seeing is itself political. Instead of opposition, there is continuity. Struggles between humans, from wage bargaining in Cuba to electoral corruption in Albania, are plainly political matters. Struggles involving both human and nonhuman lives, from the patenting of rice genes in America to the seizure by gunmen of South Korean ships fishing illegally off Somalia, are equally political. And no less political than either of these are struggles involving geophysical forces, from earthquakes triggered by groundwater extraction in Spain to the effects of pollution on the Indian monsoon. Normative analysis, blind neither to justice nor to injustice, is equally relevant at every stage. The birth of the Anthropocene is a many-sided disruption and reconfiguration of innumerable relationships within the earth system. Nothing about it should tempt us to ignore the fact that human-to-human relationships are among those being disrupted and reconfigured.

The Anthropocene does not, after all, require a turn away from the critique of sociopolitical power relations (globalization, capitalism, imperialism, and so on) toward a universal history of the human species. Instead, to understand the Anthropocene means widening the focus of sociopolitical critique and working toward *an analysis of the power relations between geophysical actors, both human and nonhuman*. It is much easier to propose this wide-angled analysis than to put it into practice, of course. But at least it does not mean abandoning the core concerns of postcolonial studies and global justice movements.

Understanding the Anthropocene depends on getting beyond interpretations of contemporary world politics that remain confined by the idea of the human (by a concern with economics, discourse, identity, and so on defined solely in human terms)—but broad interpretations of modernity that fail to take environmental factors into account are plainly inadequate, anyway. By contrast, even if political ecologists and scholars of the Anthropocene have started off on the wrong foot, they can get back on good terms as soon as both sides agree that when they talk about power relations they will sometimes mean the relations among geophysical forces, and sometimes the relations among people (which are also a type of geophysical force). They will pay attention to power relations like those determining the energy content of Hadley cells as they yield or withhold rain over water-stressed grasslands, and the balance of forces between friction and gravity in glaciers as they head toward the sea. They will recognize that contests like those cannot neatly be separated these days from other power relations, like the fluctuating influence of the Dinant Corporation over the democratic process in Honduras, the capacity of Dow Chemical to obstruct Indian corporate liability law over Bhopal, or the ability of Thai fishing peoples to defy government efforts to seize their land in the wake of a tsunami.

As I have stressed throughout this chapter, though, different conceptions of the Anthropocene have very different implications. If we want to trace the birth of the new epoch as a shifting, interwoven play of ecological powers, we will need to choose carefully the version of the Anthropocene that best enables such an analysis. No doubt there are plenty of options. If there are many interpretations of the Anthropocene, and if only some of them are politically counterproductive or philosophically incoherent, then there should still be several different ways of thinking about the Anthropocene that are stimulating and worthwhile. I do not want to be exclusive, then, but only selective, if from this point onward I focus on just a single conception of the new epoch.

We have seen that the unnerving conclusions drawn by Dipesh Chakrabarty can be set to one side by reflecting on geohistorical processes. That suggests it is well worth considering the Anthropocene specifically as a phenomenon in earth history. The rest of this book deals with a version of the Anthropocene that takes Crutzen's original proposal more literally than Crutzen himself seems to have intended. Perhaps a new epoch is indeed beginning, in the formal geological sense of that word.

THE STRATIGRAPHIC TURN

"We are assembled," wrote the stratigrapher Jan Zalasiewicz in December 2009, "to critically consider the case for a formal Anthropocene, and to make recommendations to our parent body (the Subcommission on Quaternary Stratigraphy [SQS] of the International Commission on Stratigraphy [ICS]), through them to the ICS itself, and then on to its parent body, the International Union of Geological Sciences (IUGS)." The fellowship assembled for this purpose was the new Working Group on the Anthropocene, chaired by Zalasiewicz. The significance of its remit was out of all proportion to the simplicity of its organization ("We do not have a budget," the chair reminded the group).[19] For that reason, this nested thicket of abbreviations is well worth disentangling.

The IUGS is one of the world's major scientific organizations, the professional representative body of a million earth scientists. The Commission on Stratigraphy is its largest constituent part, an organization that is essentially dedicated to finessing the one-page diagram that underpins geological science. That diagram is the International Chronostratigraphic Chart, the embodiment of the geological timescale that sets out how the history of the earth is formally divided.[20] Condensing the whole body of stratigraphic research, the chart defines, names, and dates each recognized major interval of geologic time and determines their hierarchical status and the way in which they fit inside one another. (Even putting the Anthropocene aside, disputes about where the divisions should go sometimes convulse the geological community. Stratigraphers, like poets, take naming seriously.) Definition ideally involves selecting a change—usually the appearance or disappearance of a fossil species—in a single column of rock somewhere on earth that can represent an interval's starting point. Take the Oligocene epoch of 34 to 23 million years ago, in which the mighty tropical rain forests of the postdinosaur epochs receded and the modern Antarctic ice sheets formed. The Commission on Stratigraphy defines its moment of origin as that of the formation of a layer of rock now found partway up a quarry on Mount Conero, Italy, "at the base of a greenish-grey 0.5m thick marl bed."[21]

The Commission on Stratigraphy operates some sixteen subcommissions. The correlation of stratigraphic data for the last 2.6 million years is the job of the Subcommission on Quaternary Stratigraphy. It was at the request of this last body that Zalasiewicz and the paleobiologist Mark Williams—then occupants of next-door offices at the University of Leicester—set up the unfunded Anthropocene Working Group, which in practice consists of forty academics, Crutzen among them, communicating mostly by email. "The work involved should not be onerous," they told potential participants. "However, it should be interesting, and of use to the scientific community."[22] With this gentle flattery, the idea of the Anthropocene underwent a crucial transition.

Crutzen's inspired outburst, "We're not in the Holocene anymore. We're in the . . . the . . . the Anthropocene!" had implicitly been a claim about stratigraphy, an alternative to the definitions laid down by the International Commission on Stratigraphy. But it is clear from his two foundational articles that assembling a brief for an actual revision of the Commission's great chart of earth time was by no means his priority. His own expertise was in atmospheric chemistry, and as the concept percolated through specialist literatures over the next few years, geologists themselves used the term relatively infrequently. The move toward stratigraphic formalization began with an article coauthored by twenty-one members of the Geological Society of London—not as large as the IUGS, but the oldest geological society in the world—with Zalasiewicz at their head. What had hitherto been a "vivid but informal metaphor," they wrote, could equally be scrutinized according to the "criteria used to set up new epochs." If the Anthropocene met those criteria, as seemed quite possible, the International Chronostratigraphic Chart might be amended accordingly.[23] The article marked a significant new departure from Crutzen's original idea and from all the ways in which that idea had previously been received. The working group was the result of that proposal.

Many things follow from this attempt to take the Anthropocene so literally as to incorporate it into the geological timescale, to turn it into a formal unit of geohistory. In Smith and Zeder's terms, it makes for an Anthropocene that is defined by *effects* rather than by *causes*—and by its effects not primarily for human beings but for the earth's ecological assemblage as a whole. What Zalasiewicz contemplated, in essence, was a way in which to carry out a profound displacement of the human in thinking about the Anthropocene. One would start not with human influences on the environment, not with an attribution of responsibility or blame, but with the fact of ecological change as such. So many changes, of specified magnitude, to this or that geophysical phenomenon: the sedimentation of rivers, the population distribution of phytoplankton, the acidity of oceans, the pollen content of the air. This suite of changes would then be weighed and interpreted, and their interactions reconstructed—at this stage analyzing the interspecies and intraspecies relationships of one very populous hominoid species would be crucial—in order to assess whether, and in what way, they could be said to constitute the beginning of a new epoch. Look at the earth system changes first, and let them lead you, as they undoubtedly will, to an ecology of the human species. Then pass beyond the confines of the human again, in order to grasp these processes of transformation in the terms of geologic time. This is what the stratigraphers proposed.

In order to carry out this immense conceptual displacement, a certain indirectness is needed. Instead of assessing the type and scale of present-day ecological change directly, and deciding on that basis whether the Anthropocene label is justified, one must imaginatively transfer oneself to the far future. After all, the beginning of every other epoch has been defined in distant retrospect. Stratigraphers of the Anthropocene must concentrate much less on how dramatic any given environmental change is at present than on how readily discernible it will be millions of years from now, which largely means how well traces of it will be preserved in sedimentary rocks. Some bureaucratic oddities follow from this. For instance, because marine and lacustrine deposits are usually much better preserved than those on land (which the weather erodes), stratigraphers generally focus their attention on the rock layers that build up at the bottom of oceans and lakes. Likewise, stratigraphy prioritizes hard-bodied organisms, which fossilize far more readily than soft-bodied ones, and it emphasizes the species at the base of past food pyramids over the much less abundant apex predators.

Songbirds, squid, and big cats, for those reasons, are in themselves poor stratigraphic markers for the Anthropocene. Environmentalists who are justifiably concerned about their survival might be perplexed by stratigraphers' preoccupation with fluctuations in the distribution of calcareous and siliceous marine microorganisms. But the infinitely complex interlocking of ecosystem processes means that this is much less of a problem than it first appears. Disturbances at one level will often have repercussions in another. The near-total collapse of the gargantuan Newfoundland cod population, caused by industrial overfishing, might be visible in the fossil record not directly but through changing species compositions near the bottom of the food web, where zooplankton numbers have come under pressure from booming populations of the foraging capelin that the cod once preyed upon. Levels of world GDP and of foreign direct investment—things to which Crutzen and McNeill's version of the Anthropocene gave priority—do not fossilize. By circuitous routes, however, they affect the things that do. The stratigraphic approach to defining the indicators of the Anthropocene is at once the subtlest and the most concrete of the many that have been proposed.

A second concomitant of the stratigraphic method might be equally hard to swallow at first. Stratigraphers like the dates for the beginning of new intervals to be singular, worldwide, and as exact as possible. In the present case, they have generally envisaged a formalized Anthropocene with one particular year specified as its starting point. And although a variety of candidate years have been discussed, all of them are relatively recent ones. In line with the Smith and Zeder rule that a focus on worldwide environmental effects rather than causes implies a young Anthropocene, the stratigraphers' proposed start dates all fall within the last few centuries. Opponents of the Anthropocene have generally denounced this way of dating the new epoch as a gross simplification, one that neglects both the deep roots of ecological change and the gradual and geographically variable nature of industrial modernization. But perhaps those opponents have not yet asked themselves whether they really wish to accuse geologists of being intellectually uncomfortable with the idea of long-drawn-out change. As we will see, the stratigraphic version of the Anthropocene does not remotely imply a one-step model of environmental transformation.

I have argued in this chapter that several different versions of the Anthropocene are possible. Still other uses of the term will no doubt emerge in the future. But the turn toward stratigraphic formalization provides the most fertile way so far to interpret this epoch-making play of power relations among human and nonhuman forces. The stratigraphic approach reinvents the Anthropocene by giving it a place amid the complex mosaic of the geological timescale. What remains is to elucidate the stratigraphic conception of the new epoch and to use that conception as a way of changing the debate about the politics of the Anthropocene. Taking our cue from the geologists will mean paying attention to the lineaments of a history that goes back much further than 1784, even if a date like that eventually proves to be a good candidate for the epoch's formal starting point. To interpret the Anthropocene stratigraphically means placing it in the context of geological time. This is the method that will help us see why contemporary environmental problems have started to load the pages of daily newspapers with references to the events of a hundred thousand or three million years ago.

CHAPTER THREE

Geology of the Future

In a nutshell, my argument so far has been as follows. There are several versions of the Anthropocene, so if you want to talk about it, you have to specify the version that you mean. The environmental crisis of the present day plunges those living through it back into deep time, and one version of the Anthropocene—the one that takes literally the possibility of defining it as a new geological epoch—provides a way of coming to terms with that disturbing encounter with the distant past. In other words, thinking through the implications of adding the Anthropocene epoch to the geological timescale is a way of locating the current crisis within earth history.

That means that there is much to be learned from the activities of the ICS Anthropocene Working Group, the volunteer corps of Jan Zalasiewicz and his colleagues. For one thing, the significance of the name *Anthropocene* shifts when it is understood as a stratigraphic term. If the new epoch's name works in the same way as the names of other units in the geological timescale, then it is free from many of the unwelcome overtones of which it has been accused. The stratigraphers' research program begins not with any assertions about who is responsible for environmental disaster, but with a leap of the imagination. What traces might the events of the recent past leave on the earth thousands or millions of years into the future, they have asked, and what does that tell us about where contemporary changes belong in the long narrative of earth history?

The formalization of the Anthropocene epoch within the geological timescale will require more than just predictions, however. It will depend on evidence that one can already recognize substantial, well-defined changes to the composition of the earth that have come about in recent decades or centuries. The key to the stratigraphic Anthropocene is a tightly specified starting point: a "base," in the jargon. Various candidates for the base of the Anthropocene have been proposed, and the strength or weakness of those candidacies will determine whether the new epoch can be formally ratified. More importantly, it is through the evaluation of those candidates that the stratigraphic study of the Anthropocene intersects with, and shines a light on, the environmental and economic history of the last five centuries. Each of the main candidates offers us a distinctive way in which to understand the place of the modern crisis within deep time.

THE NAME OF THE ANTHROPOCENE

The Anthropocene is a brilliantly provocative label for the new epoch. If the epoch is formalized in the geological timescale, it will certainly be given the name that supposedly came to Paul Crutzen in a flash, rather than being called *the Econocene, the Cosmoscene,* or anything else. And this pointed name is undoubtedly one of the main reasons for the concept's viral spread. It condenses into a single word a gripping and intuitive story about human influences on the planet. It works, on the most basic level, as a kind of shock tactic: the planet has changed so much that many scientists believe we have entered a whole new geological epoch! Or rather, its message seems at first blush to be: *humankind has changed the planet so much that it has created* a new geologic epoch. The Anthropocene has often been considered a counter-Copernican idea. Whereas Copernicus displaced human beings from their privileged place at the geographical center of the universe, the Anthropocene, in this line of thinking, puts human beings back at the center of the physical world.

The theme of a transition between geological epochs would never have commanded so much attention, so far beyond the profession of stratigraphy, had it not brought human activity to the foreground in such a compelling way. On the other hand, it is clear that the mere choice of name is also one of the main reasons why so many critics have rushed to declare that the Anthropocene tout court is an unacceptably universalist and totalizing idea. For its opponents, as we have seen, naming the new epoch "the Anthropocene" is tantamount to dismissing all differences between groups of human beings and thereby blaming the whole world's population indiscriminately for environmental catastrophe, in a way that can only lead to the adoption of counterproductive and antiegalitarian solutions. Evidently, a great deal depends upon what exactly the relationship is between humankind, the *anthropos*, on the one hand, and the Anthropocene epoch on the other.[1]

Zalasiewicz and his collaborators have been happy to persevere with Crutzen's name for the epoch, even though their conception of the Anthropocene is subtly but fundamentally different from his own. We need to understand the sense in which the word *Anthropocene* is being used when it is employed as a part of the specialized terminology of stratigraphic science. What happens to the relationship between the Anthropocene and the *anthropos* if the new word is taken literally as the name of a geological interval? If we consider how the names of geological intervals are usually formed, we will see that the stratigraphic approach implies a much less exclusive relationship between the two than many interpreters have assumed. Speaking of the Anthropocene certainly does not mean, in this context, that human beings are singlehandedly the creators of a new phase in earth's history.

The Carboniferous period gets its name from the Latin *carbō* (charcoal, or coal) because one distinctive feature of the time between 360 and

300 million years ago was the accumulation of massive coal deposits. The Cretaceous period was so called because many large and striking formations of chalk (in Latin, *crēta*), such as the White Cliffs of Dover, date to the last 80 million years before the extinction of the land dinosaurs. Coal deposition was an important part of the Carboniferous, and chalk accumulation was important to the Cretaceous. But it would make no sense to imagine that the whole Carboniferous period was an inherently coal-centered one. Nor did the volcanoes or the ankylosaurs of the Cretaceous have any kind of chalky essence about them. In both periods, ecological processes went on in their own ways: tectonic plates shoved and buckled; the sun burned off vapor from the seas; insects buzzed between stands of fern. Meanwhile dead trees were crushed underground into coal, and the calcium carbonate shells of marine creatures built up in chalky layers. At certain times the latter two processes left such heavy traces in the rock record that nineteenth-century geologists used them to provide names for slices of earth's history. But this Victorian nomenclature does not make them any more special than that. Many other deposits of coal and chalk were laid down outside the periods that bear their names, just as the Cryogenian period (from the Greek κρύοσ, "frost") was a remarkably cold interval, but by no means the only one in which glaciers formed. In each of these cases, conceptual priority lies with the interval itself, considered as a span of time with certain distinguishing characteristics, more or less dissimilar to the times before and after. Only afterward was some characteristic feature of the interval pressed into service to supply a name for the whole thing.

The point is, of course, that all this applies to the Anthropocene too. The belief that the *anthropos* must be the essence or metaphysical centerpiece of the Anthropocene is without foundation. When the word is used as the name of a stratigraphic unit, it does not imply that the Anthropocene is the "epoch of humanity" in the sense that it contains nothing other than human agency, or in the sense that all the rest of the world is subordinated to human dominion. If the physical world has changed so much in recent centuries that a fresh geological epoch can be said to have begun, then a name like *Anthropocene* seems appropriate, because human activities have been extremely prominent among those changes. Nonetheless, these ecological reconfigurations have let other players besides the *anthropos* exert themselves too, unbidden by humans and not reducible to their desires.

The Anthropocene includes volcanic eruptions and undersea landslips as well as mountaintop removal mines. *Leptinotarsa decemlineata*, the Colorado potato beetle, first evolved to eat potatoes in the nineteenth century, and it has eaten them in great numbers ever since, having developed resistance to an array of pesticides. Chlorine atoms climbed to the stratosphere in CFCs, each unraveling thousands of ozone molecules there, decades before anyone below conceived of such processes. Humans made those transformations possible, but they can hardly be said to have had them under control. Colorado beetles and chlorine atoms, like gray squirrels and kudzu, are among the powers of the Anthropocene. They are just as real as human beings, and just as capable of exerting themselves within their own spheres of influence. The idea of the Anthropocene puts all of them on the same ontological plane. Humanity, in this epoch, does not absorb or command some merely passive nature, issuing orders from a central throne to the dull physical substance that surrounds it. On the contrary, human societies are only the most vigorous and distinctive among an irreducibly various array of altered forces. Thus the turn to stratigraphy helps one assess the status of the *anthropos* within the Anthropocene in a realistic and levelheaded way.

In this light, the claim that the Anthropocene is an innately dualistic concept loses its force. Jason W. Moore, the Anthropocene's sharpest and most accomplished critic, sees the whole concept as resting on an unacknowledged binary whereby "humanity" is seen as interacting with a "nature" from which it remains essentially separate and independent. It is certainly true that some individual writers on the new epoch have been snared by such dualist habits of thought, but that tells us little about the Anthropocene in its stratigraphic sense. Compare the Devonian period, the predecessor to the Carboniferous. Unless that period implies that reality is fundamentally divided into two, and only two, parts—with the pleasant English county of Devon forming one half of reality, and everything else in existence, from the rings of Saturn to mental states in Penzance, crammed into the other half—there is no reason to think that the Anthropocene rests on a philosophical division between humans and nature.

All of this makes a difference to the practical definition of the Anthropocene. As we have seen, proponents of an "early Anthropocene" define the beginning of the new age by focusing, as Bruce Smith and Melinda Zeder put it, "on cause rather than effect, on human behaviour." For them, the difference between the world of the Anthropocene and all the rest of the earth's history is that the former is marked by human presence. Of course there is nothing to stop them using the word in that way, but as cogent as Smith and Zeder's version of the Anthropocene is on its own terms, it is not a stratigraphic concept. Geological units mark collective, worldwide transformations, rather than private milestones in the career of one species. Members of the Anthropocene Working Group clarified that point in the course of rebutting an otherwise forgettable attack on the concept: "We do not believe that it is necessary to seek a 'boundary stratigraphic marker' that reflects the time 'since anthropogenic change began,'" they wrote. "The issue here is not the presence or absence of human traces in strata. It is whether Earth's stratigraphic record—and the processes that shape it—have changed sufficiently to make a new unit justifiable and useful and, if so, to seek the most effectively traceable boundary horizon for it."[2]

The stratigraphers' concern is with the suite of geophysical changes that constitute the latest discontinuity in the history of the earth, not with the beginnings of human influence on the planet as such. For that reason, they always envisage a relatively recent starting date for the Anthropocene: they regard it as beginning only with the radical changes that the workings of the world have undergone in the last few centuries. They see no reason why the epoch should embrace the entire time during which humans have had a significant impact on the state of the planet, especially given that "significant" seems almost impossible to define in that context. *The Anthropocene* is not a suitable name for the new epoch merely because human beings are an ecological force in it. It has long been understood that humans were also one of the ecological forces that molded the Pleistocene epoch, and by the same token, beavers, bacteria, and bryophytes are still ecological forces in the Anthropocene. The point is that transformations that may well be great enough to justify the declaration of a new epoch have taken place within the last three centuries, and that it seems appropriate to christen that epoch "the Anthropocene" because human agency is outstandingly prominent among its novel biogeochemical assemblages.

What we now think of as the Anthropocene could even be a unit of geologic time that endures long after humans' deliberate influence on the planet has largely been lost. The Working Group's members have been clear about this mordant implication of their thinking as well. As some of them put it: "It may be misleading . . . to think of the Anthropocene just as the 'human epoch.' The key factor is the level of geologically significant global change, with humans currently happening to be the primary drivers: future, potentially yet more pronounced change may be primarily driven by Earth system feedbacks such as methane release, and yet would still clearly be part of the same phenomenon."[3]

In that grim scenario, the epoch that has now begun would ultimately prove to be characterized much more by the destabilization of methane clathrates than by intentional human modification of ecosystems. Human activity could be brutally curtailed by a species-wide population collapse in the early part of the epoch, whereas methane outgassing could leave a much deeper geological trace via a process of rapid greenhouse warming, ice sheet melting, and sea level rise—a series of events that has taken place several times in the past. In geological terms, the two parts of the process would form a single act in the drama of geohistory, regardless of the fact that one was consciously driven by human beings and the other was not. And even in that scenario, *the Anthropocene* could still be a reasonable name for the epoch, if anyone is around to bestow it. Humans would at least have initiated the chain of events that characterized the epoch, although under those circumstances no one could imagine that the name meant they were its masters.

The Anthropocene of the geologists is not an anthropocentric concept, nor one that separates humankind from the rest of nature. The name *Anthropocene* describes the most distinctive aspect of the new epoch, not its single essence. And if the Anthropocene does not put any human beings at the ontological center of the world, then it is certainly not guilty of putting all of them there in a lump. That is, if the new epoch is not a dualistic concept, then the further accusation that it reduces the diverse human populations of the world to an undifferentiated mass, all collectively responsible for current environmental despoliation, also loses its power. The geological Anthropocene is neither universalist nor technocratic, and neither deterministic nor antipolitical. Rather than designating a general human footprint on the natural world, it implies only a network of evolutionary developments and ecological interactions.

GEOLOGISTS FROM SPACE

The stratigraphic conception of the Anthropocene begins with a thought experiment. "Let us admit, though eccentric it might be, the supposition that a strange intelligence should come to study the Earth in a day when human progeny . . . has disappeared completely." That strange intelligence would be able to understand the world "only by putting in all his calculations this new element, human spirit. . . . So that future geologist, wishing to study our epoch's geology, would end up narrating the history of human intelligence."[4] Thus wrote one of the forefathers of the Anthropocene, Antonio Stoppani, in the 1870s. In recent years, the thought experiment that he envisaged has actually been conducted in the writings of Jan Zalasiewicz and his colleagues.[5] The defining signals of the stratigraphic Anthropocene are not the rising levels of GDP and international tourism emphasized by Crutzen and his coauthors, or even the shares of land and biological productivity presently co-opted by humans that I described at the end of the first chapter. Instead, they are all those changes to the earth that might be discernible in the distant future, because of the way in which they alter the layers of sediment and snow that will be stacked and compressed into rocks and ice sheets.

Suppose that human civilization were to fade from existence over the coming years, so that its final geological footprint was similar to that which it would leave behind today. What would happen if aliens were to land on the earth a hundred million years from now and study it with geological techniques like those of the present? Zalasiewicz has argued that even at that distant time the planet would still harbor the records of human technology, in a distinctive layer of rock that would in effect constitute the lower bound of the Anthropocene epoch.

In a hundred million years' time, most of the sedimentary rock strata being laid down today will have disappeared irretrievably, having been scoured by ocean currents, eroded in the air, or dipped into the earth's molten interior. The remainder will mostly be buried deep underground or underwater, and virtually inaccessible. Nonetheless, some will happen to be at or near the surface, just as some rocks formed from hundred-million-year-old sediments are visible today. And just as those mid-Cretaceous rocks make possible the reconstruction of part of a colorful narrative when they are read in juxtaposition to those above and below them, so too will the rocks of the human interval. They will not appear utterly different in kind to every other layer of stone on the planet's surface, but nor will they be dully identical to all the rocks around them. Instead, they will record one unique moment among many others in the course of geohistory.

As Zalasiewicz imagines it, the alien geologists of a hundred million years' time would only gradually have their attention drawn toward the strata currently being laid down. Tracing the history of the planet on which they had landed, they would reconstruct the contrasting fossil assemblages of successive ages and grow particularly interested in the turning points between them. Most of the major turning points would be mass extinctions, and as we have seen, modern times are yet to witness a full-scale mass extinction. Nonetheless, the rocks laid down above those of the present day would hold arrangements of fossils that were radically different from those below. The difference would arise mainly from a geographic redistribution of species unprecedented in the record of complex life. The effects of that redistribution would be permanent, since future evolution would have taken place on the basis of the new arrangements.

Introduced flora now make up close to half of the plant species found on remote islands like New Zealand and Hawaii, and more than 20 percent even in places like Britain, Canada, and New England.[6] Introduced species dominate overwhelmingly the fauna of Australia and the Americas, where they constitute (for instance) up to 99 percent of the biomass in San Francisco Bay. The alien geologists would be able to discern many of these introductions directly, when fossilized skeletons, leaves, footprints, or pollen showed up far outside the range recorded in earlier rocks. The workings of fossilization would mean that the clearest such records would be those of abundant shell-forming shallow-water organisms, typically dispersed worldwide in ships' ballast water. But many other, subtler traces of the relocations could also be found. In the wetlands of the Great Plains, for instance, the Eurasian common reed has displaced native willows, and the fine-grained, silica-rich sediments that have built up among the stands of reeds may provide the raw material for distinctive sedimentary rock layers. New species have arisen through hybridization with interlopers, or in response to them, like the five moth species that have evolved to feed on bananas in Hawaii. Others have been altered, like the soapberry bugs in Australia that have evolved longer beaks in order to feed on invasive balloon vines.[7] The strata of the present day would carry the evidence of innumerable near-simultaneous ecological modifications like these. Standing out among them would be the geological sites that bore witness to the unceremonious simplification of ecosystems that is associated with the most destructive invasions: by cane toads, chestnut blight, wolf snails, fire ants, rats, rabbits, chytrid fungus, zebra mussels, Nile perch, Japanese knotweed, and the like.

This paleontological merry-go-round would very probably be synchronous with the leveling-off of many of the Brobdingnagian limestone plateaux that are formed by coral reefs. Corals can live only in sunlit upper waters, and rising seas at a time of miscellaneous hardship—the world's corals are being overheated, acidified, poisoned with industrial waste, muddied by sediment from deforested shores, blanketed by seaweed fed on agricultural runoff, and dynamited for fish—currently appear set to drown them. (More than half of the world's reef-building corals have already been lost, a decline unprecedented for tens of millions of years.) The flat-topped limestone mountains left behind should be starkly visible to the aliens.

Once the aliens had identified the turning point—the biohorizon—at which the transplantations and the coral wipeout began, they would have good reason to examine closely the strata of the present day. They would find there an episode of global warming (reflected by changes in the proportions of different isotopes of oxygen in ocean sediments) that might mark the start of a 130,000-year hiatus within a spell of cold climate that had begun about 2.6 million years earlier.[8] They would probably identify a rise in sea level, a phenomenon that always transforms the distribution of sediments beneath wide expanses of the sea, not just the newly flooded land. Recent sea level rise has been negligible by geological standards, but the destruction of the

West Antarctic ice sheet is now considered inevitable.[9] If the Greenland sheet follows, that is already a rise of twelve meters, which would have potentially long-term visibility.

The pollen grains trapped in the strata would reveal further distinctive qualities: signs of rapid deforestation despite the warming climate, combined with exceptional worldwide abundance for just a handful of plant breeds. (These would be the domesticated staple crop grasses; the global spread of corn, which produces an abundance of pollen, would be especially noticeable.) The shells of marine microorganisms would contain an unusual mix of carbon isotopes, having taken up the carbon released by burning fossil fuels. Despite the comparative patchiness of the fossil record of large animals, the skeletons of domesticated vertebrates should certainly be found. There would be a strange paucity of limestone in ocean sediments, replaced by a layer of clay: evidence of acidification. Some evidence of modern-day iron and especially steel production—a combined total of 15 billion metric tons so far—might have been preserved, and several of the formerly rare minerals now found in enormous quantities in cement, bricks, and ceramics would probably have survived robustly. Deep boreholes, mines, and underground nuclear test sites and storage facilities might well remain virtually intact, and even collapsed mines would still exist as channels of brecciated, or fragmented, rock.[10]

This thought experiment is the basis for the claim that the Anthropocene may legitimately be introduced as a new geological unit. The distinctive character of contemporary strata can already be specified with considerable confidence, even though events still to come will affect the way that they would appear to far-future observers. The particular nature of the changes that mark the beginning of the Anthropocene will be unique, as is always the case. But the *types* of changes will mostly fall into categories with which stratigraphic analysis is perfectly familiar. The lower bound of the Anthropocene will be marked by worldwide evidence of things for which geologists have a well-honed jargon: a climate transition, a marine transgression (i.e., sea level rise), a "reef gap," an episode of mineral diversification, a carbon isotope excursion, and distinctive bioturbation (the spoil heaps and bore-holes, analogous to animal burrows). The boundary will be marked especially clearly as the base of numerous assemblage zones (the species relocations), and in palynological (pollen) analysis. The proponents of the stratigraphic Anthropocene have always claimed that they are not asking for any special treatment in the recognition of the new epoch. They argue that—although of course it is an unusual case in many respects—the Anthropocene epoch meets the regular criteria normally required for the ratification of geochronological units.

THE HUMAN STRATUM

The case for formalizing the Anthropocene in the geological timescale rests on changes to the composition of rock strata that would take place world-wide. Few of the changes that I have described would necessarily reveal much about the primate species whose activities link them all together. In principle, the alien geologists could identify the present day as a significant turning point while learning very little about the part played in it by Stoppani's "new element," *Homo sapiens*. Recognition of the Anthropocene does not—this point can hardly be stressed too heavily or too often—mean asserting that the whole world is now subordinated to human agency. It just means that a suite of changes significant even in a deep-time perspective is taking place in planetary systems.

That said, Zalasiewicz has argued that even in a hundred million years' time, clear records of human existence probably will remain on earth. Having become interested in present-day strata, he supposes, the hypothetical aliens would track the sediments laid down in today's oceans into shallower and shallower water. Then, by carefully tracing present-day shorelines, they could hunt down what Zalasiewicz calls the "Urban Stratum": the remnants of coastal cities. (Urban areas now cover 3 percent of the land surface, concentrated in preservation-friendly coastal and deltaic locations.) The cities' state of preservation would depend on how rapidly sea level rise had sunk them beneath the erosional surf zone, but if fossilized dinosaur bones can survive a hundred million years underground then so, a fortiori, can artifacts of ceramic and lead. The aliens would be able to excavate the rubble of concrete buildings, now turned "decalcified and crumbly"; car parts crushed into "irregular patches of iron oxides and sulphides"; lines of "softened brick"; opaque fragments of glass.[11] They would even, Zalasiewicz hazards, be able to reconstruct the gross anatomy of the species who built the cities. Humans are extraordinarily numerous for a megafaunal species, and formal burial multiplies skeletons' chances of survival as well as preserving them in tellingly ordered rows.

Geologists who proceeded that far would evidently have recognized the traces of intelligent beings. But they would probably be able to deduce much less about the conscious life of the species that they studied than about its characteristic behavior. Zalasiewicz shows how humankind could stand revealed in the geologic record as a single, herding, migratory species, omnivorous in diet (much of its food web could be reconstructed) and technological in habits (mines would give evidence of fossil fuel use). The age structure of the fossil assemblages would suggest that juveniles were cared for. Conversely, some skeletons might reveal evidence of deliberate killing, so the aliens would have no reason to envisage the species as Gerda Roelvink's undifferentiated "universal and transhistorical collective," but could guess at intergroup violence and war. But the aliens would be limited, Zalasiewicz says, to analyzing human beings in "broad ecological terms" like these. "It is hard to think how the normal workings of geology and taphonomy [fossilization] can capture anything that one might describe as embodying the essence of humanity," he observes, given that the works of Mozart and Schubert, Shakespeare and Goethe, Michelangelo and Rodin, will be definitively irrecoverable in a hundred million years.[12] That is true enough in itself, but what should be added is that the aliens' lack of access to the ideas bound up in human books and paintings might bring with it certain benefits as well as losses.

As Zalasiewicz says, the aliens' viewpoint would necessarily be an "ecological" and nonhumanist one. Theirs would be an interpretation of the human species derived from the shape and intensity of its material interactions with other beings and forces—coal, rice, coral, nitrogen, iron—and not from its inward self-imagining. That interpretation would without doubt be incomplete, but it could be thought of, too, as a rigorous and disenchanted one. Because, despite what Zalasiewicz implies, even unhindered access to the masterpieces of the Western canon does not put one in touch with "the essence of humanity." Even Schubert and Shakespeare were products of their time and place, not mediators of a timeless human spirit. Such a spirit or essence finally has no more reality for us than for the alien geologists of the distant future. What the aliens would lack is just the *illusion* of a transcendent human essence. It is worth trying to see the world somewhat in the same way as those imaginary aliens, then. Doing so could make a difference to the ecological politics and criticism of our own time.

The thought experiment that gives rise to the stratigraphic Anthropocene tries to understand the present by imagining its geological traces. Far from planting the essence of humanity at the dead center of the world, it humbles all such humanist pretensions. The alien perspective of the far future would be one in which plastics, grasses, humans, plankton, and carbon dioxide molecules were all bundled together. Those geologists' object of inquiry would necessarily be the entire ecosystem that had collectively given rise to the rock record that they studied. In their eyes, each factor in the planetary system would be neither more nor less real than every other. Ecological thinking that learns from the aliens' example could never be a matter of making "man" and "nature" grind perpetually against one another like Chinese medicine balls.

The planetary forces addressed by this ecological thinking would be those that actually exist at the time of the Anthropocene's birth. Everyone has heard the claim that green politics is special because is concerned with the future: that environmentalists, unlike others, look beyond the next electoral cycle to worry about the interests of future generations and about the state of the planet that we will hand on to our grandchildren. But that claim is largely bogus. Secular politicians of every stripe likewise promise their voters a better future and a more prosperous life for their children. A genuinely emancipatory green politics would mistrust this rhetoric of kinship and patrimony, and would go in fear of sacrificing the interests of the living and impoverished to the interests of the unborn children of the West. The attitude to future time implied by the stratigraphers of the Anthropocene is exemplary in this respect. They do indeed look to the future. But they look much, much further into the future than most environmental thinkers, and they do so precisely as a way of looking back on the present. Their concern is not the well-being of the planet in a hundred million years, but the current state of the biosphere, which they want to situate in its deep-time context so as to understand it better. Imagining how the earth will look in remote ages becomes a way in which to focus on concrete realities that are immediately at hand.

The stratigraphers of the Anthropocene thus take a view of deep time exactly opposite to the Olympian pretensions that, in chapter 1, we saw shared by Colin Tudge, David Brower, and Matt Ridley. For their different reasons, the latter three all at times have declared great tracts of history to be much of a muchness and insisted on disregarding the challenge of living amid the particular circumstances of the present. The idea of the Anthropocene, by contrast, is a way of sharpening one's focus on the dangers and transformations that are peculiar to the contemporary world. Those dangers stand out most clearly against the backdrop of geologic time. That is why it is crucial to remember that the Anthropocene Working Group's object of study is not the Anthropocene epoch as a whole but only the very earliest fraction of this new unit of time. Whereas its beginning is all around us, the later course of the epoch (which we may imagine lasting millions of years into the future) is almost completely unpredictable. It might be dominated by an intense methane-driven thermal maximum and a mass extinction, as in the Working Group's gloomy scenario described above. Equally, it might happen to be a more equable time than many other geologic epochs. But in either case, *Homo sapiens* may well disappear from the fossil record somewhere in its lower reaches. So there is no particular need for anyone to worry about the distant upper stages of the Anthropocene (let alone the "post-Anthropocene"). What humankind has to deal with is only the pressing reality of the epoch's lower boundary: its birth pangs.

GOLDEN SPIKES

Looking back over a distance of a hundred million years, geologists are hard pressed to distinguish dates more precisely than to the nearest hundred thousand years or so. To Zalasiewicz's imagined aliens, the manufacture of Aurignacian flint tools and the building of the Burj Khalifa in Dubai would look more or less simultaneous. In their eyes, the beginning—the "base"—of the new interval might just be a single human event layer. But if they could somehow reconstruct human history as precisely as present-day scholars are able to, when exactly would they place the start of the epoch?

We saw in the previous chapter that debates about what the Anthropocene *is* often take the form of debates about when the Anthropocene *begins*. Everyone agrees that the earth's ongoing change of state is a drawn-out process taking place over centuries, at least, with antecedents that go back much further still. Nonetheless, the boundaries between the various components of the geological timescale always have to be fixed somewhere, or the study of earth history would descend into a morass of terminological confusion. That means that if the proposed new interval is to be represented on the International Chronostratigraphic Chart, then some particular date must be picked out to represent the moment of change from the Holocene epoch to the Anthropocene. Given how short the relevant timescales are, it has seemed to most workers that a single year ought to be chosen as the turning point. Trying to decide on the most apposite year might seem like splitting hairs when we compare it to the perspective of Zalasiewicz's aliens. But in fact the debate about the starting date deserves all the energy that is being expended upon it. The attempt to assign a precise date to the new epoch is what brings the concept of the stratigraphic Anthropocene up against the specifics of global environmental history.

The base of the Anthropocene could be defined in one of two ways. The simpler option would be to assign to it some appropriate numerical age, so that the epoch would begin in, say, 1784 or 1950. This method—the selection of a Global Standard Stratigraphic Age (GSSA)—is the one used for defining early intervals in earth's history, before the development of complex life, because of the paucity of paleontological evidence. Stratigraphers of the Anthropocene face some similar challenges for the opposite reason, a super-abundance of data, while they also have the advantage of being able to choose a date based on precisely known historical records. For both reasons, there is much to be said for defining the Anthropocene directly via a GSSA.

In the main, however, an alternative method is preferred, one based on real-world reference points. With this approach, some specific change in a sedimentary archive—like the beginning of that green-gray marl bed in an Italian quarry—is chosen to represent the transition between geological intervals and is named as a Global Boundary Stratotype Section and Point, or GSSP. Practicalities allowing, a golden metal marker may then be hammered into the relevant layer of rock: for that reason, GSSPs are often referred to as "golden spikes." The interval is defined by the change in the rock record where the golden spike is fixed. The timing of the interval is then derived by investigating the date of this tangible alteration in the composition of the earth. Geologists are currently completing the process of fixing all other geochronological boundaries for the last half a billion years on the basis of GSSPs. If the Anthropocene too is to be based on a golden spike, in line with this preferred approach, then a signal that represents its beginning must be picked out from among the multitude of recent transformations in the makeup of the planet.

The perfect stratigraphic marker for demarcating a unit of earth history would be one that was found all over the world, and that was unambiguously visible to every trained observer. It would appear everywhere in the stratigraphic record at the same moment, and it would be possible to date that moment precisely. It would be preserved within continuously deposited layers of rock, so that the preceding and succeeding environments could be thoroughly compared, and for extra security it would come with a cluster of independent auxiliary markers. It would appear likely to survive indefinitely into the future.

Not surprisingly, perfect markers for stratigraphic boundaries do not exist. Most boundaries are associated with the first or last known appearance of particular fossil species. The fossils chosen for this purpose are always common, widespread, and well-recognizable ones. Even so, imprecision is inevitable. The earliest and latest known specimens are almost certainly not the first or last members of the species ever to have existed. The species must have originated in one part of the world before spreading more widely, a process that may have taken many thousands of years, and the dates assigned to its appearance in any given location may have uncertainty ranges of hundreds of thousands of years. The species will have flourished only in a limited range of habitats, only a small proportion of those habitats will have been preserved so as to be accessible for study, and a still smaller proportion of them will actually have been studied. The standard of preservation even in those sites will be variable; in particular, the amount and type of contextual evidence preserved alongside the reference species will fluctuate enormously. The most readily accessible exposures will often be susceptible to being eroded away over time.

Given all those commonplace uncertainties, prospective markers of the Anthropocene should not be held to a standard of perfection that other golden spikes do not reach. It is not an argument against the idea of the Anthropocene to say that there is no unambiguous marker of its beginning or even a single candidate clearly superior to all the others: the representative of its lower bound will inevitably have to be selected through a process of debate. The rules of that debate, however, can at least be specified in advance. Established stratigraphic procedures provide the criteria by which proposed indicators of the new epoch can be judged.

Some thoughtful observers doubt that any prospective marker of the Anthropocene will be able to meet the usual criteria for the ratification of geological epochs. In this line of thinking, there is insufficient justification (at least so far) for introducing a new epoch to the geological record, and the Anthropocene ought not to be formalized by the institutions of stratigraphic science. These skeptics are not persuaded that the geological formalization of the Anthropocene would fulfill the most fundamental requirement, that of usefulness. After all, the history of recent centuries is recorded in infinitely more detail in written records than in layers of ocean sediment. The relevant sedimentary layers might be too thin and too ambiguous as yet to provide a robust basis for a new epoch. The justification for the Anthropocene involves predictions about how well the traces of current humans will be preserved, and how clearly they will stand out to future observers, but it is at least conceivable that "another extended interval of voluminous flood basalts or another large asteroid impact [might yet] overwhelm any sedimentological/stratigraphic record of human activities."[13]

Furthermore, without diminishing any of the scientific evidence indicating the gravity of anthropogenic change, one might wonder whether a clear geological break from the preceding Holocene epoch can yet be seen, especially remembering that the Holocene, too, has always been marked by intensifying human impacts. The current wave of extinctions—fewer than 1 percent of all species, as we have seen—is not yet epochal in scale; there has yet to be a geologically significant rise in sea levels; the biggest effects of climate change are still to come; and the worst effects of many regional crises of pollution, overfishing, and habitat destruction might still be ameliorated if sufficient political will is brought to bear. Conversely, if current environmental pressures appear set to further intensify and converge, that too might be an argument against a precipitate declaration of the Anthropocene. Demographic trends suggest that the human population will not reach a plateau before the second half of the present century, at somewhere on the order of ten billion souls, and there seems every reason to predict a sharpening of conflicts related to climate, energy, water, pollution, and land use into at least the 2020s and 2030s. If most aspects of the world crisis currently have such strong momentum behind them, it may be wise to wait before ratifying the new epoch. A compelling candidate for the base of the Anthropocene might lie unforeseeably in the future: an exchange of nuclear weapons, say, or a transformative program of geoengineering.

In short, it may be premature at best to inscribe the Anthropocene in the Chronostratigraphic Chart. In the absence of institutional ratification, the word could continue to be used in an informal and relatively loose fashion, and this flexible approach might be the most pragmatic and illuminating option.

Without being persuaded by it, I can see the force of that argument. Equally possible is a subtle variation on the skeptical viewpoint that ultimately has quite different implications. Suppose you agree that it is too early to declare the beginning of the Anthropocene, because the most appropriate boundary for a new geological epoch may perhaps lie in the future, possibly centuries away. You might (without inconsistency) hold that view while also thinking that current evidence already justifies the claim that the world is irreversibly locked into the process of undergoing an epoch-level transition. That is, you might believe that the transition to the next epoch is not yet complete while also agreeing with Paul Crutzen that the world has already changed so much that the Holocene cannot outlast the present crisis in any meaningful sense.

According to this variation on the skeptical view, either the Anthropocene has already begun, or the Anthropocene *will* begin. If the Anthropocene has not yet arrived, the reason for that can only be that it will arrive differently—still more forcefully, or still more decisively—at some future time. In that case, the world is now precisely in the marginal state between two geological epochs. Perhaps the generations living today are not exactly witnessing a process of transition, in the sense of being able to measure and track it. Perhaps instead (and at the risk of sounding paradoxical) what they are witnessing is the very impossibility of witnessing that transition. They are living through a time so disruptive, and a change so great, that it will be possible to make sense of it, and hence to reliably date the birth of the Anthropocene, only in retrospect. Even the most astute contemporary observers are seeing the new epoch in such tight close-up that they can scarcely see it at all. The implication is that the border between the Holocene and the Anthropocene is not just somewhere close at hand but is itself the defining feature of the contemporary world. Until we can look back and say that the birth of the Anthropocene has been completed, that its beginning has come to an end, we seem to be neither in one geologic epoch nor in the other, but in a fissure between the two.

That argument, like the full-scale skeptical one, involves rejecting the stratigraphic formalization of the Anthropocene, at least for now. If the argument sounds appealing, it is a reminder that the possibility of ratification by the IUGS is not the most important thing about the stratigraphic version of the Anthropocene. (The most important thing is the opportunity that it provides to grasp the environmental crisis by putting it in the context of deep time.) But despite all the difficulties, a strong case is steadily being assembled for a formalized Anthropocene based either on a calendrical date (a GSSA) or on a golden spike in the existing sedimentary record (a GSSP). Much of that case is contained in a landmark publication by the Geological Society of London, a volume called *A Stratigraphical Basis for the Anthropocene* assembled by a group of scientists led by the geologist Colin Waters, the secretary of the Anthropocene Working Group and one of its key organizers. Assessing the various candidates for the golden spike means looking at how the idea of the Anthropocene runs up against the last five hundred years of world environmental history.

THE CANDIDATES (1): CORN, TUNNELS, COAL

The Anthropocene in its stratigraphic sense will not have a premedieval starting point. We have seen William Ruddiman and others propose Anthropocenes that began many thousands or even millions of years ago, having used the word in some other sense. Lake beds, deltas, and pollen assemblages record millennia-old evidence of deforestation, soil erosion, and soil modification, sometimes over wide areas. A layer of elevated lead concentrations in Greenland ice cores bears witness to Carthaginian and Roman mining in southern Spain.[14] But changes like these do not constitute an epoch-level transformation in the earth system as a whole. They are scattered widely in time, their long-term visibility is often limited, and they nearly all took place on a less than continental scale. All plausible epoch boundaries fall well within the last millennium.

Indeed, the editors of *A Stratigraphical Basis for the Anthropocene* are inclined to rule out any starting points preceding the Industrial Revolution. That might be too hasty, however. The Columbian exchange—the ecological fusion of Afro-Eurasia and the Americas across the Atlantic and, later, the Pacific—presents candidates for the Anthropocene's golden spike that deserve more consideration than they have yet received.[15] Thinking only in terms of progressively increasing human mastery over the nonhuman world makes it harder to see what took place. In the Columbian exchange, human agency took on a slightly different role, one to which it is especially well adapted: humans acted as a conduit. The voyages of Christopher Columbus and the *marinheiros* provided a means for Afro-Eurasian and American land biota to become entangled with and act upon one another. Their ships reestablished a low-latitude connection between the two continental clusters that—as we will see below—had been separated by oceans for millions of years. An event that is novel on a million-year scale is exactly the sort of thing that might be expected to mark an epoch-level boundary, and the circulation of crops, weeds, animals, and diseases that it enabled did indeed have immense consequences for the stratigraphic record, as well as for the course of human history.

The effects were most drastic in the Americas. American landscapes (as well as those of the Old World) were already heavily influenced by human societies: in the Andes and Amazonia, Mesoamerica, southwestern North America, and the Eastern Woodlands by more or less intensive food-producing cultures; and elsewhere by hunter-gatherer chiefdoms and bands that held down populations of keystone prey species and managed the land with fire. European colonizers, however, introduced radically new environmental regimes with remarkable speed. Human populations crashed in disease epidemics, war, and forced-labor enterprises; the continents and their seas were stripped of beavers, otters, whales, fur seals, and cod; soils were pillaged for cash-crop monocultures of sugar, cotton, coffee, and tobacco; forests were consumed in timber and shipbuilding, or regrew prolifically over the ruins of destroyed civilizations; weeds and feral horses, cattle, and pigs spread far beyond the settler zones; and older hydrological and fire regimes were transformed by dams, irrigation, and antiburning laws. Meanwhile, American crops spread through Afro-Eurasia. The biggest uptake of New World staples, and perhaps the most rapid, took place in the most populous and advanced region of the world: China, where the adoption of corn (maize), sweet potatoes, and peanuts helped the population to more than double during the Ming dynasty (and then keep rising, a boom that in time brought with it chronic soil erosion and geopolitical vulnerability). Corn and cassava became African staples. Potatoes, complemented by corn, fed Europe's peasantry.

In their different ways, all these changes left behind enduring geological traces that might serve for stratigraphic purposes. If the Anthropocene was taken to begin in the era of the Columbian exchange, 1492 would be the obvious date for a definition by GSSA (the assignation of a fixed starting date). A GSSP (a real-world reference point, or golden spike) for the Anthropocene could be based on an assemblage zone arising from the first appearance of corn pollen in Eurasian sediments, or from the first appearance of Old World domestic fauna—horses might be the most suitable species—in the archaeological record of the Americas.[16] That seems apt, given that species relocations are such an important part of the Anthropocene's geological signal.

Nonetheless, the geological imprints of the Columbian exchange are still markedly "time transgressive," spreading from region to region over the course of centuries. It would be a mistake to place too much emphasis on 1492 as the year in which the modern world was born. The way in which the Western European states exploited the Americas was not a given, but was in part the outcome of their economic transformation during the previous century, and of several external adventures that had their own stratigraphic implications, including a dress rehearsal in the invasion of the Canary Islands. It is even more important to note that the colonization of the New World did not by any means produce a European-centered world economy and ecology straight away. Instead, its first effect—still an extremely significant one—was to incorporate the Americas into the peripheral or hinterland region of a global trade network that remained centered on the Indian Ocean and East Asia. Even with the expropriation of the ecological resources and precious metals of the New World, it took until about the end of the eighteenth century for any European states to achieve parity in development with the most advanced regions of China and India.

William Ruddiman and his sympathizers hold that American reforestation produced a measurable decline in atmospheric CO_2, contributing to the Little Ice Age, but aside from that controversial hypothesis the effects of the transoceanic exchanges were largely confined to the biosphere, as opposed to the atmosphere or hydrosphere. And although the ecological changes that took place were almost certainly the most dramatic since the end of the last glacial period more than ten thousand years earlier, it might be hard to argue that they were of an epochal magnitude in a stratigraphic sense of the word. A starting point in the fifteenth or sixteenth century would be at the outer limits of possible datings for the new epoch. On the whole, the golden spike is probably better placed somewhere further down the road.

At a minimum, however, the socioecological upheavals of this period make it an essential precursor to the birth of the Anthropocene. The best approach might be to regard the later fifteenth century not as the start of the Anthropocene per se but as the beginning of the transitional phase between the Holocene and its successor.

The fifteenth and sixteenth centuries saw the emergence of the European capitalist regime of globalizing commodity chains and of capitalism's concomitant exploitation and degradation of successive regional ecologies all around the world. To students of world history, this period has often seemed to constitute the decisive rupture that marks the beginning of the modern world system. Thus, when Jason Moore, the most eloquent opponent of the concept of "the Anthropocene" (in fact, only of Crutzen's first-draft Anthropocene), presented his "Capitalocene" as an alternative, the single most obvious difference between the two was that the Capitalocene was supposed to begin in the "long" sixteenth century, whereas Crutzen's Anthropocene was originally dated to the late eighteenth century. The idea of a transition into the Anthropocene that has been under way since the second half of the fifteenth century would bring the theme of the Anthropocene disarmingly close to Moore's own way of seeing.[17] In Moore's analysis, the crisis of the medieval world produced a coalescence of interests between Europe's monarchical state bureaucracies, postfeudal seigneurs, and mercantile city-states in favor of capitalist geographical expansion, and this in turn initiated an enterprise that required the rapid commodification and depletion of the ecological wealth of the Atlantic islands, West Africa, the Caribbean, Brazil, the Indian Ocean spice islands, and so on. That coalescence of interests was the spark; the Holocene-Anthropocene transition was the explosion. Geologists speak of the "end-Permian event" or the "end-Triassic event" at the close of older geological intervals. A geological perspective invites us to conceive of the postmedieval world system in its entirety as what we could similarly call the *end-Holocene event*—an event that is still playing itself out.

The Anthropocene Working Group has given far more attention to possible starting points for the Anthropocene on the other side of the seventeenth-century world crisis, ever since Crutzen's initial proposal that the epoch might be best dated to 1784 and one of James Watt's patents on steam engine design. The much-discussed idea of linking the base of the Anthropocene to the British Industrial Revolution certainly has its merits. However, a somewhat narrow understanding of industrialization has so far constrained most of the debate about placing a GSSA or GSSP in the long nineteenth century. According to this model, the Industrial Revolution was initiated by the invention and deployment of new technologies; it began in Britain and then rippled outward, starting to affect other nations only once their own industrialization began. That diffusionist hypothesis has in turn given rise to concerns that a late-eighteenth- or early-nineteenth-century golden spike would be hopelessly Eurocentric. But understanding the Industrial Revolution in a different way, as an event in world economic history, can dispel those fears of Eurocentrism. Britain's textile and iron-smelting sectors were indeed pioneers of industrial production, but their rapid evolution was only one element of the general restructuring of a trade system across all five major continents, a system that metabolized South American silver, Caribbean sugar, North American cotton, African slaves, and the consumer goods that flowed into and out of the advanced population centers of northwest Europe, China, and India.

At the heart of the economic and demographic factors that drove Britain's pioneer industrialization were the high price of labor there relative to energy, and its colonial strong-arm tactics. England had retained a relatively high-wage economy since the Black Death—when fewer peasants meant stronger bargaining powers—partly because its non-rice agriculture made feeding workers costlier than it was for its Asian peers. Its readily accessible northeastern coalfields made energy cheap. The imbalance arising from that coincidence made it uniquely worthwhile there to invest in technology that substituted machine power for labor time. It was especially worth doing so in the cotton sector, because British manufacturers could extract cheap raw cotton from slave-worked plantations in the Caribbean and the American South and export finished textiles to the New World at a handsome profit. There still resulted little in the way of breakthrough conceptual innovations, by James Watt or anyone else (in fact, the heroic age of the European physical sciences was the seventeenth century, not the eighteenth). Instead, however, a stream of technological refinements and efficiency gains succeeded one another until British manufactured goods had become cheaper than their rivals' worldwide. Thus Britain's industrialization had what climatologists call a teleconnection to deindustrialization in India.

For most of the eighteenth century, China and India retained their long-standing competitive advantages over the polities and markets of Europe. But in the first half of the nineteenth century, after the destruction of the Indian princely states by the East India Company (exploiting the eighteenth-century fission of the Mughal empire), the subcontinent was transformed. Previously an advanced competitor economy—India, not Britain, had been the world's leading cotton manufacturer—it was reduced to serving as another source of raw materials, and another destination, unprotected by trade tariffs, for British manufactured goods. China, too, was broken open in the middle of the century, in a cycle of economic decay, peasant rebellion, and Western incursion, preceded by environmental disasters and spearheaded by the armed drug dealers of the British empire. By 1820, Britain's empire incorporated more than a quarter of the world's population. That colonial hinterland was fundamental to the conditions—a strong central state and ample capital, as well as high wages and vast open markets—that impelled it toward mechanized and coal-fueled production.

In short, the geological traces of industrialization in early nineteenth-century Europe (the Low Countries, as well as Britain) are shaped fundamentally by the manufacture of goods for export to other continents, goods that displaced non-European manufacturing in the aftermath of military assaults and colonization. A golden spike in the early nineteenth century, even one based on ecological changes that were local to western Europe, could have real legitimacy. It would reflect a worldwide economic, demographic, and political restructuring, rather than an exclusively European phenomenon.

Whether a robust stratigraphic marker can be found anywhere in the long nineteenth century is another matter, however, especially given the absence of any starting point for the period that is as comparatively clear-cut as the *marinheiros'* ocean navigations. If coal-based industrialization is seen as a defining characteristic of the age, then the contamination of lacustrine, peat, or coastal sediments by the fine particles released during fossil fuel combustion is an apt candidate for the golden spike (although localized pollution from coal burning can be identified much earlier, and the scale of the contamination would accelerate more sharply at a later date). Another candidate is the coal residue, or clinker, tossed overboard from nineteenth-century steamships, which created distinctive geological traces in the sediment-accumulating, and previously relatively untouched, sea-bed. There were some substantial population rises—the population of Britain and Ireland doubled in the sixty years to 1841—with commensurate expansions in agricultural production, livestock holdings, and urban areas, and a further increase in the spread of invasive species. To contemporaries, deforestation was the most visible and alarming sign of environmental change in Eurasia, with the loss of trees already a significant concern in China (especially the south) and Japan as well as large parts of Europe—Britain, the Low Countries, the German lands, and the Mediterranean—by 1800. All those changes had their effects on fossil assemblages, pollen records, and the sedimentation of rivers, lakes, and deltas, some of which might provide sources for Anthropocene stratotypes.

The growth and increasing architectural complexity of cities make for an especially interesting candidacy. One of the most fundamental divisions in the geological timescale, that between the Proterozoic and Phanerozoic eons, is marked by the burrows of a little-understood wormlike creature, which stratigraphers treat as a metonym for the emergence of newly complex animal behavior. The base of the Anthropocene could be marked in the same way, by the increasing complexity of humans' burrowing in their urban habitats. Thus, a group of researchers led by Mark Williams has suggested a possible golden spike located in one of the original stations of London's underground Metropolitan Railway, dating the birth of the Anthropocene to 1863.[18]

The nineteenth century differs from the early modern period in the spread of lasting anthropogenic traces from the biosphere into the atmosphere. That opens up the possibility of a golden spike defined by reference to ice sheets, speleothems (especially stalagmites), or the chemistry of the shells of marine microorganisms. We could, after all, endorse Crutzen's initial idea of defining the Anthropocene by the increasing atmospheric concentration of CO_2, or more directly by shifts in carbon isotope ratios in the bodies of marine organisms. Changes in atmospheric carbon have the advantage of being unarguably global, but the downside is that, on a year-to-year scale, CO_2 concentrations change smoothly rather than discontinuously, so the golden spike would have to reflect an essentially symbolic milestone, such as atmospheric CO_2 levels surpassing three hundred parts per million soon after 1900.

A final option provides a useful reminder that there is no reason why the golden spike for the Anthropocene has to have a human origin at all. In April 1815, one of the largest volcanic eruptions in recent millennia issued from Mount Tambora in Indonesia, reflected in a deep red tone in the sunset paintings of J. M. W. Turner and afflicting most of the world's population with three years of abysmal weather and catastrophic crop failures. The eruption left sulfate peaks in the ice at both poles and in tropical glaciers, as well as a layer of tephra particles stretching for thousands of kilometers (although not worldwide).[19] Having taken place during the Congress of Vienna and Napoleon's Hundred Days, the eruption coincides precisely with the beginning of the stable reactionary order that underpinned stage 1 of Crutzen and McNeill's two-stage Anthropocene. If it is chosen as the Anthropocene's golden spike it would usefully press home the point that the new epoch cannot possibly be characterized by the abolition of nonhuman influences on the geologic record. But perhaps, ultimately, no single nineteenth-century candidate really stands out clearly from the rest. Potential dates for the base of the Anthropocene are scattered rather loosely through the long nineteenth century; many of the potential markers are in some sense localized rather than global; and in most cases their long-term geological visibility must be in some doubt.

THE CANDIDATES (II): CONCRETE, LEAD, PLUTONIUM

With potential golden spikes from the fifteenth century to the early twentieth looking plausible but not compelling, stratigraphers have increasingly focused on the prospect of a still more recent date for the birth of the Anthropocene. Potential starting points within the last generation have scarcely been discussed, because the sedimentary layers would in that case be desperately thin. But the mid-twentieth century is another matter. The idea of dating the Anthropocene to somewhere around 1950 is currently in the ascendancy. Doing so would acknowledge just how recent the arrival of the modern world-ecological order really is, and how dramatically the human-influenced changes to the planetary system have multiplied in the last seven decades. The human population in 1950 was about 2.5 billion. In 2011, it reached 7 billion. Total economic activity increased elevenfold over the same period, as global average GDP growth per capita accelerated after the body blows of the world wars and the Great Depression, from less than 1 percent annually in 1913–50 to nearly 3 percent in 1950–73. The techniques and intensities of the environmental exploitation that underlay that growth developed beyond all recognition. (At the outset of World War II, for instance, even British farms used ten times as many horses as they did tractors.)[20] Perhaps it was only now that an epoch-level shift in the world's workings really gathered pace.

One significant component of the boom was that other advanced nations caught up to the levels of affluence, and of investment in capital-intensive technology, that the United States had already achieved by 1945. Another was the effect, in the same countries, of progressive economic planning and of a working social compact between capital and labor. But very much as with the Industrial Revolution, it is possible to take too narrow a view of the *trente glorieuses* of the mid-twentieth century. It might be imagined that the new economic order began in the most developed countries and affected the rest of the world only once it rippled out from them to the regions that lagged behind. In reality, since worldwide trade integration was by now a long-established fact, a more global and systemic perspective on the economic and ecological structure of the Great Acceleration is required.

The Cold War partition of the world into U.S. and Soviet spheres provided a new way to manage the terms on which the global South participated in the world economy. That economic periphery was essential in enabling the core manufacturing economies of the North to function as they did, with their ever-higher wage levels and ecological throughput. In the underdeveloped nations, regulatory obstacles to the expropriation of communal assets were weak. That made them the indispensable suppliers of cheap industrial raw materials and biological assets (including oil above all), of low-skilled labor reserves, and of dumping grounds for pollution and surplus production. That is why—as John McNeill's environmental history of the twentieth century shows in compelling detail—the sites of ultrarapid ecological alteration and degradation in the postwar era were scattered all around the world, rather than being confined to the neighborhood of the most profitable industrial zones.[21] The wretched of the earth were not excluded from the Great Acceleration but an integral part of it, albeit on terms set by the economic elite. Very much as with the Industrial Revolution, the implication is that potential stratigraphic markers in this period that are immediately derived from the rich world's industry, technology, urbanization, or (it's worth underlining) military and geopolitical strategy need not be considered Eurocentric or NATO-centric. Rather, they reflect structural changes that affected states and populations all around the world.

The biggest difficulty with placing the Anthropocene's golden spike in the period after World War II is the brevity of the timescale. The base of the Anthropocene would be associated with unconsolidated surface sediments, teetering on the upper edge of the geologic record and still highly vulnerable to disturbance. In the deep oceans, where disturbances are fewer, only tissue-thin layers of Anthropocene-epoch material would as yet have accumulated. Boundaries set some centuries or even millennia earlier would still face similar problems, however, and the counterbalancing advantages of the postwar period are the increased size and long-term visibility of alterations in the earth system, and their tighter synchronization around the world.

The many plausible candidates for a golden spike in these decades all have their own benefits and downsides. The exponential postwar growth of megacities provides a potential marker, as does the construction of a world-encircling paved road network (now greater in extent than the iridium-rich clay layer, formed by the dinosaur-killing Chicxulub impact, that divides the Cretaceous period from the Paleogene). The golden spike could pick out alterations in sediment dynamics caused by unprecedented dam-building, earthworks, irrigation, and bottom-trawling. It could reflect the intensification of mining and drilling activities, or the larger carbon isotope excursion resulting from accelerated fossil fuel burning. A group of candidates referring to offshore drilling, deep-sea litter, and disturbance of the ocean bed has the advantage that transformations beyond the continental shelf began from something more like a standing start in the mid-twentieth century, but the disadvantage that because sediment accumulation rates in the deep ocean are usually very slow, it is hard for finely datable layers to form. Potential golden spikes based on mineral diversification likewise cluster in the postwar period, given the huge multiplication then in the production of cement, plastics, iron, steel, bricks, glass, ceramics, and so on: 98 percent of aluminum production has taken place since 1950, for instance.

Many of the strongest contenders have a biological basis, among them the devastation of coral reefs. Species relocations had already changed the world, as we have seen, but the paleontologist Anthony Barnosky argues that "it is likely that 1950 would closely approximate the time at which mixes of native and non-native species became widespread in the sedimentary record." Again, small marine animals and pollen-heavy plants—barnacles, mussels, clams, grasses—are the introduced species most significant for stratigraphic purposes. Barnosky also proposes a possible Anthropocene lineage zone: "palaeontologists of the future" should be able to recognize the first appearance of supersweet strains of corn, which were bred for the first time at midcentury.[22] Microfossils, the fundamental currency of most stratigraphic analysis, are better suited to abundance-zone approaches, which pay attention to changes in the population sizes of species in a given location rather than to their first occurrences. The midcentury saw a general proliferation of microfauna and microflora "tolerant of eutrophication, hypoxia, metal pollution, water acidification [and] salinity change."[23] Foraminifera, beloved by paleontologists for their well-fossilized shells; ostracods, tiny crustaceans; diatoms, photosynthetic algae; and the cysts produced by whip-propelled dinoflagellates all sensitively record the pollution, salinization, eutrophication, and warming of bodies of water. The dominant influences on these microorganisms are usually local to their catchment areas. But selecting one such local incident as representative of the global transition between epochs is less problematic when near-simultaneous pollution episodes are taking place around the world.

Even snowbound lakes in the Rockies and the high north, far from the sources of pollution, show well-synchronized alterations after 1950. Rainfall and snowfall have carried the effects of the recent doubling of the world's reactive nitrogen supply all the way to those remote lakes. Specifically, that contamination can be identified via a decrease in the proportion of a particular nitrogen isotope, nitrogen-15, within the total quantity of nitrogen in lake-bed sediment samples. That decrease is a strong candidate for the golden spike, again dating to the postwar era: the use of nitrogen fertilizer increased ninefold from 1960 to 2000.[24] Other globally dispersed chemicals provide comparable reference points. The concentration of lead in Greenland ice peaked in the 1960s at two hundred times background levels (it has since declined with the switch to unleaded gasoline). Antimony concentrations tell a similar story, except that they are continuing to increase. The burning of coal and fuel oil at temperatures above a thousand degrees Celsius creates a unique particulate form of black carbon, and such particles have scattered into lake beds, peatlands, and ice around the world. Human industry appears to be their only source, so they represent directly a principal driver of global change, and their rate of accumulation increased sharply in the middle of the twentieth century. They have excellent preservation potential.[25]

Finally, and most persuasively of all, there is the possibility of a definition based on nuclear explosions. Two papers by members of the Anthropocene Working Group have set out what look, for now, like the leading candidates for definitions by a GSSA and a GSSP. If a boundary is chosen based on a calendrical age, or GSSA, it could be set with unimprovable specificity on July 16, 1945, "at 05:29:21 Mountain War Time (± 2 s)."[26] This is the moment of the Manhattan Project's first nuclear weapon test, Trinity: white light in the predawn New Mexico desert. The bombing of Hiroshima followed three weeks later. The Trinity test is as clear and discrete a marker as any that could be imagined. It was an objectively new physical phenomenon: the first runaway nuclear reaction on the surface of the earth. Its significance was plainly global rather than localized, even if American victory in the Pacific was already inevitable by the summer of 1945 and the bomb's main strategic effect lay in the way that it shaped the Cold War world. A GSSA set in mid-July 1945 might not help dispel concerns that proponents of the Anthropocene are too preoccupied with high technology and with events in the rich countries of the West. But its military quality should surely dispatch the objection that the concept of the Anthropocene involves thinking of all humankind as an undivided mass. The best thing of all about this choice of a base would be its lucid simplicity.

The Trinity test cooked the white sands of the desert below it into a novel kind of rock, Trinitite. Its geological impact did not extend beyond the test site itself, however, so it does not supply a potential golden spike. There is much to be said for the straightforwardness of a definition by a fixed chronological age, but basing the Anthropocene on a GSSA would put the new epoch problematically at odds with the current preference for establishing the geological intervals of the last half a billion years (and, ideally, older ones as well) on the solid basis of tangible physical markers in the constitution of the earth.[27]

An alternative is to formulate a GSSP, or golden spike, by looking to the radionuclide fallout from nuclear tests. The fission bombs of the 1940s, horrifying as they seemed at the time, were tiny compared to the fusion or thermonuclear weapons that were tested from 1952 onward.[28] For six years, test shots by the United States, Soviet Union, and United Kingdom in the tropical Pacific and Kazakhstan spread increasing volumes of radioactive fallout all around the world. A moratorium on nuclear testing from 1958 to 1961 brought a dip in soil radioactivity levels, but that truce progressively unraveled until, in October 1961, the Soviet Union set off the largest of all human-made explosions, the Tsar Bomba, at its Arctic island test site. There followed two years of intensive open-air testing, conducted in a spirit of geostrategic posturing during the era of the Cuban missile crisis rather than simply for research purposes, and a corresponding surge in plutonium fallout. In 1963 the Limited Test Ban Treaty required that explosions take place underground, as public fears about fallout—and the superpowers' desire to restrict access to the nuclear club—grew stronger. The fallout peak came in 1964, just after the signing of the treaty. A steep decline followed, down to indiscernible concentrations in the early 1980s.

Open-air thermonuclear tests dispersed radioactive isotopes of cesium, strontium, americium, plutonium, and carbon as high as the stratosphere before they settled down into soils and sediments worldwide. Of these elements, plutonium is the most significant for stratigraphic purposes (indeed, it is already sometimes used by environmental scientists in the dating of sedimentary records). The half-life of plutonium-239 is 24,110 years; it binds tightly to soil particles, and its preanthropogenic concentration in the earth's crust was extremely low. With the use of accelerator mass spectrometry techniques, the twentieth-century spike in plutonium levels will remain detectable in sediment samples across much of the earth for a very long time: not the next hundred million years, but at least the next hundred thousand.

The quality of the plutonium record will vary between sedimentary archives. Worms, plowing, and rainfall will blur the fallout pattern in ordinary soils. A less compromised trace will survive in marine and coastal sediments, stalagmites, and coral cores. There will be a still crisper one in ice sheets and glaciers, and in lake beds that are largely free from sediment run-off: notably the beds of those dazzling lakes that form in the craters of extinct volcanoes, which are filled from the air rather than from watersheds. The crater lakes with the best-preserved records will be those with the least oxygen and hence the fewest living things in their lower reaches, minimizing disturbance of the smooth layers of sediment. Those located in the mid-latitudes of the Northern Hemisphere, where nuclear fallout was heaviest, will have an extra advantage. The golden spike could be placed there: in a sediment core drilled from beneath such a lake, and preserved in some university or museum, a sliver of compressed mud would serve as the birth certificate of the Anthropocene.

If plutonium-239 radionuclides are chosen as the marker of the base of the Anthropocene, it might seem logical to date the new epoch to the year of maximum fallout, 1964, making it coincide with the beginning of the Swinging Sixties. Standard practice, however, is to place stratigraphic boundaries at the start rather than at the peak of the isotope anomaly that defines them, since the former is usually more easily identified (even though in this case one might stretch a point, since the 1964 peak is very pronounced).[29] That would place the birth of the Anthropocene in 1952, at the beginning of the rise in worldwide plutonium concentrations. Instead of originating in the year of Brezhnev and Beatlemania, the Anthropocene would have been born amid postwar high modernism and decolonization: in the year of John Cage's *4'33"*, Le Corbusier's Unité d'Habitation, and Ralph Ellison's *Invisible Man;* of Nasser's coup in Egypt, civil disobedience by the ANC in South Africa, and the Mau Mau rising in Kenya.

To underscore a basic point: this does not mean that plutonium-239 fallout from the Ivy Mike test shot in November 1952 pushed the world from one geological epoch into another. On its own, the fallout's ecological significance is not remotely great enough for that. But the radionuclides that it dispersed could be selected as the emblem of a suite of co-occurring upheavals in the biosphere, atmosphere, hydrosphere, and pedosphere: changes that, taken as a whole, constitute the end of the Holocene and the start of the Anthropocene. A still more basic point: it *should* go without saying that fixing the base of the Anthropocene in 1964 would not imply that *With the Beatles* was released into a world that had seen no significant human impacts on the earth system, and would not imply that by the time *Rubber Soul* came out the planet was an exclusively human construct. Instead, 1952 or 1964 would be adopted as the indicator of a passage between geological epochs (not at all a passage between nonhuman and human worlds) that has been hastening toward completion ever since the fifteenth century, and that still has further to run. The plutonium-239 spike would be the synecdoche for an ongoing end-Holocene crisis that may be identified with the emergence of the capitalist world order.

We need to take a step beyond Dipesh Chakrabarty's opposition between the story of capital and the story of geology (or "globalization and global warming") in order to see capitalist modernity in this new light, as the very process of transition between the Holocene and the Anthropocene. The plutonium fallout record is the best available reference point—albeit no more than that—for this transitional phase. Chronologically, the fallout spike might in the end prove to lie roughly midway between the voyage of Columbus's *Santa María* and the final disappearance of the West Antarctic ice sheet. Wherever the boundary layer is finally placed, it should always be remembered that other ways of dating the epoch had been possible. But let's say, for argument's sake, that the Anthropocene epoch began in 1952.

+ + + +

It's time to take stock. The pair of chapters that conclude this book are quite different from the preceding three, so I will finish this first part of the book by boiling down the version of the Anthropocene that I have developed here to five key maxims or principles. I have said that the word *Anthropocene* has a variety of possible and legitimate meanings; for this reason, it is always worth being clear about the way in which it is being used. Here is a reminder of the meaning that it has in this book.

Firstly, the Anthropocene, in this context, is a neocatastrophist term. It belongs within the late twentieth century's paradigm shift in understandings of how the earth system works. Planetary systems that were once regarded as stately and slow-moving now appear to be characterized by unstable feedback loops; biological and climatic processes are deeply interlinked, and both are susceptible to abrupt changes of state. That explains how an epoch-level transition can take place in just a few centuries, and how a single species can be preeminent among the drivers of change. It follows that the birth of the Anthropocene is not a fall from Eden or an unnatural acceleration of the old, slow, boring regimes of geohistory. Instead, it means stumbling into another of the fissures that crisscross deep time like crazy paving.

Secondly, it is not the case that the special properties of human beings are the ultimate cause or origin of the Anthropocene. The Anthropocene is not "the interval in which humans play a significant role in shaping the earth system" (what can "significant" possibly mean there?). Instead, the epoch can be defined only through its differences from other epochs: the Anthropocene begins when the characteristic conditions of the Holocene no longer exist. The new epoch emerges through a particular configuration of ecological agencies that has, among other things, temporarily multiplied the effectiveness of *some* human actors. The thought experiment about the alien geologists reminds us that those humans are just one among innumerable forces that shape the fossil record: it cautions against a dualism that would split humans off from "nature." One consequence is that even the death of the human species would not necessarily mean that the world was any-where near the end of the Anthropocene, or that the "post-Anthropocene" had begun.

Thirdly, the strongest force at work in the birth of the Anthropocene is best thought of not as humankind per se but as human *societies*. An extrava-gantly disproportionate degree of world-making power attaches to only a minority of the world's people. No adequate account of the Anthropocene can begin by lumping all human beings together, as if they are all equally the cause of global change and all identically vulnerable to its effects—not even if one subsequently bolts on an acknowledgment of socially differentiated responsibilities. The reason is that the idea of the Anthropocene is all about a new configuration of biogeological struggles, conflicts, and incursions, and that (in line with the second principle) there is no justification for excluding human-versus-human struggles from the list of those conflicts that are disrupted and reworked. The stratigraphic version of the Anthro-pocene offers no excuse for neglecting the role of class relations or transna-tional inequality in environmental change, because it is precisely focused on the changing power relations throughout ecological assemblages of all kinds.

Fourthly, the politically salient issue is the time of transition into the Anthropocene, not the new epoch as such, because it is almost idle to worry about what the world will look like in a hundred thousand years. The idea of the Anthropocene does not give any succor to those who would adopt a pose of godlike detachment from the mere transient life of the present day. Instead, stratigraphic science provides a model of earnest engagement with the particular characteristics of singular moments in the history of the earth. The stratigraphic Anthropocene places the present crisis in its deep-time context only so as to let its distinctive features be seen more clearly and to help in assessing its significance. It has the potential to foster a deeper sense of entanglement in immediate historical circumstances, rather than an indifferent acceptance of the fact that nothing lasts forever. Who could be more fully immersed in geohistory than the witnesses to the birth of a new epoch?

Fifthly, the reason why the idea of the Anthropocene can make a telling contribution to contemporary politics is that geological time has already become a political factor. Atmospheric CO_2 concentrations have reached levels not seen for three million years; Arctic temperatures seem to be at their highest for more than a hundred thousand years; and the sixth great extinction in the history of complex life may be getting under way. It is news like this that makes the Anthropocene something more than an exercise in stratigraphic whimsy or the latest piece of fashionable academic jargon. Even granted that the word may be used in multiple senses, it is nonetheless a waste to use *the Anthropocene* merely as a modish way of referring to environmental degradation in some ill-defined slice of the recent past. The term is urgently needed for a different purpose: as a way to help us get a grasp on the fact that green politics now has to confront the role that human societies play in deep time itself.

These five maxims sum up my argument so far. Elaborating the idea of the Anthropocene epoch is a way of opening a window onto deep time, to help in taking the measure of the planetary crisis. What remains is to actually take a look through that opened window. The remaining two chapters of this book describe what the history of the earth looks like from the vantage point of the Anthropocene. This sketch of earth history will introduce some new complications. It will uncover a deep ambivalence in the very idea of a geological epoch; it will mean acknowledging that there is nothing new, after all, about humans being entangled in deep time; and it will involve not only intervals far longer than epochs but also some other geological time units that are surprisingly human-size. It should also, however, illuminate everything that has gone before. We have seen Don McKay describe how earlier epochs seem to run backward from the new arrival "like rungs on a ladder," leading the mind's eye swiftly away to a time before human existence. We need to know a little about those rungs on the ladder before we can really understand the Anthropocene of the stratigraphers.

CHAPTER FOUR

The Rungs on the Ladder

The Subcommission on Quaternary Stratigraphy's most controversial working group treats the Anthropocene as an epoch in the strict, geological sense. That means that the Anthropocene Working Group's concept makes sense only in relation to the system that organizes and classifies units of geological time. Seeing what is distinctive about the Anthropocene means knowing what differentiates it from the epoch that preceded it, the Holocene. But before we can properly understand the relationship between those two epochs we must also know something about how they relate to the ones that came before them. Moreover, for stratigraphers, an epoch is a unit of time that belongs within a nested hierarchy of other units, like a monarch's reign within a dynasty. Any given epoch must be understood with reference to the geologic *period* of which it forms a part. That period in turn can be appreciated only in relation both to earlier periods and to the still larger unit within which it, too, belongs. To make sense of these interlocking units of time, it is necessary to gaze far back into the history of the earth.

Perhaps in principle this widening of the frame of reference cannot stop anywhere. The back story of the Anthropocene runs the risk of turning into a bewildering dream, one in which mountain ranges flare and subside, continents nose their way through long-extinct oceans, and the rungs on McKay's ladder lead straight down into the swirl of interstellar dust out of which the earth itself was formed. This chapter and the next do not look all the way back to that swirl of dust. But in a pair of narrative time lines—the first one, in this chapter, on a million-year timescale; the second zooming in to a thousand-year scale—they lay out the scene of deep time within which human societies are now performing their histrionic role. The two chapters are a primer for making sense of the numbers bandied about in the stories from Mauna Loa and the West Antarctic ice sheet.

Certainly, by no means every detail about long-dead species or the geography of Gondwanaland in the story that follows is a prerequisite for sensible assessments of present-day environmental problems. But the only way of showing just how haphazard and chancy—how *historical*—the course of geohistory has actually been is to tell the rambunctious tale of the larger units inside which the Anthropocene belongs. The birth of the epoch gives us a standpoint on deep time, and looking back on deep time is what gives the new epoch its significance. Sampling the catastrophes, the painful destructiveness, and the innumerable new beginnings that make up earth history is what will let us understand that the Anthropocene does not stand apart from the drama of deep time but instead belongs, for good and for ill, right inside its tempestuous course.

FROM THE SNOWBALL TO THE ASTEROID (640 TO 66 MILLION YEARS AGO)

The first thing to do is to draw a line somewhere. Defining the prehistory of the Anthropocene very generously indeed, so as to reduce to a minimum the feeling of picking up the story partway through, we might trace it back as far as 640 million years ago. At an inch to a year, that corresponds roughly to the distance from New York to Sydney. The vast spans of time preceding that date have a complex and interesting history in their own right, of course, but they do not concern us here. Six hundred forty million years of context represents the outer limit of practical usefulness in setting the scene for the present ecological crisis. It is enough to make sense of even the most far-reaching stories about the environment that might appear in tomorrow's newspaper.[1]

The first 100 million years since that date can serve as a kind of prologue to the story of the Anthropocene. That prologue begins in the earth's wintertime. As its historians like to put it, 640 million years ago the planet was a snowball. Its land and oceans were very largely—some think almost entirely—encased in deep ice. Seasonal temperature variations were extreme, so that at the height of summer the tropics warmed up at least to temperatures like those of the modern Arctic. Average global temperatures, however, may have been closer to minus fifty degrees Celsius. Despite the unthinkable cold, simple life-forms existed in this world: largely single-celled bacteria and archaea, but also tiny multicellular algae and the earliest, microscopic animals, including sponges and the ancestors of jellyfish. In the most intense snowball scenario, they must have clustered in a handful of refuges such as patches of melting ice in the tropical summer, unfrozen spots above undersea volcanoes, and areas of ice-free water kept clear by chance currents. Carbon dioxide levels in the atmosphere were immensely high, and rising: the ice shield meant that the CO_2 steadily being emitted by volcanoes had had no opportunity to leave the atmosphere by being absorbed back into the oceans.

Not long after this starting point, the resultant greenhouse effect finally reached a tipping point of almost inconceivable magnitude. The ice melted, and melted; an increase in average global temperature in the region of one hundred degrees Celsius has been proposed.[2] The CO_2 levels meant that the rain now falling in this furnace was fiercely acidic. But life, again, clung on. Indeed, it soon began to do more than cling. The retreat of the ice meant that microorganisms adaptable enough to negotiate the fury of the climate were

The Rungs on the Ladder

now free to expand into boundless uninhabited territories. And quite soon after the end of the snowball, the world's first visible-size creatures appeared. Precisely how soon is debatable, but in any case the gap is small enough to have made many scientists wonder about a causal relationship. The new creatures were the Ediacaran organisms: soft-bodied, sea-dwelling digesters of bacteria, some over a meter long and perhaps floppily capable of a little movement, with wondrously varied body plans in the shapes of cords, fronds, lemons, and quills. It is not yet understood whether the Ediacarans are a profound root of the tree of life, or whether they were beings equally remote from plants and from animals, who died away without issue.

The Ediacarans' world, however, still falls just before the story of the Anthropocene proper. At the top level of the geological timescale, the history of the earth is split into four *eons*. The fourth and current eon is called the Phanerozoic: the eon, that is, of "visible life." The diagram near the start of this book shows the Phanerozoic eon and its main divisions. And as that diagram indicates, the Anthropocene—considered stratigraphically—is in essence a candidate to become the latest sub-sub-subdivision of the Phanerozoic. The story of the new epoch really begins with the start of the eon that embraces it, 541 million years ago.

So what is the Phanerozoic? The most general possible account of it involves the persistence of complex life, punctuated by occasional massive extinctions. The preconception most likely to hinder an understanding of the eon is what Stephen Jay Gould called faith in the "cone of increasing diversity," the belief that "life begins with the restricted and simple, and progresses ever upward to more and more and, by implication, better and better."[3] That is, one might imagine that Phanerozoic life would have begun with a handful of simple, small, and primitive species, all much alike, and would then have ramified slowly and progressively, as new organisms perpetually improved on their predecessors by growing more sophisticated and better adapted to their surroundings. Humans, of course, would be at the top of the pile. Gould himself saw in the fossil record the converse process, a dramatic *winnowing* of anatomical variety early in the eon, another hypothesis that is generally out of favor now. But, remarkably enough, all the major broadscale categories of organisms in earth's biosphere (all the major "phyla") do seem to have evolved in the very earliest part of the eon. Witnesses against the progressivist assumption include creatures like the anomalocaridids. Sometimes more than a meter long, with nutcracker mouths and sophisticated compound eyes, they probably made their living as strong-swimming apex predators. They first appeared right at the start of the eon. Living things have not shown any general tendency to become bigger, faster, more complex, or cleverer in the 500 million years since the anomalocaridids first flourished.

On the other hand, new tricks have intermittently been added to the biosphere's repertoire as the eon has gone on, sometimes enabling it to grow in overall size. Life did not develop ways to cope with dry land until well after the first 100 million years of the eon. Calcium-carbonate-producing plankton in the open ocean are a keystone of the biogeochemistry of the modern earth system, but over 200 million years of the eon passed before they are known to have first appeared. Nonetheless, the evolutionary history of the Phanerozoic is demonstrably much more a matter of contingency, of geohistory, than of steady upward progress. The configurations of living things simply changed when the ecosystems that they were a part of changed (by growing hotter, wetter, more dominated by a particular predator, more saline, richer in nitrogen, or anything else). Those changes were not teleological, nor predictable. Recognizing this characteristic of the Phanerozoic is a prerequisite for making sense of the particular sub-sub-subdivision of the eon with which we are concerned.

The phenomenon that began the eon 541 million years ago was the so-called Cambrian explosion (in the sense of an explosively rapid diversification of life-forms). The oceans stood high in the early part of the eon, and rich marine ecosystems proliferated on the shallow beds of the broad, warm seaways that ran into the continental interiors. Some of the Cambrian organisms, unlike the Ediacarans, had hard bodies: their skeletons and sharp teeth indicate predator-prey relationships, like those between anomalocaridids and the trilobites that they hunted. The burgeoning marine life of the first 100 million years of the Phanerozoic also saw the emergence of creatures from starfish to sea snails and feathery-limbed sea lilies, coral reefs that formed huge atolls and enclosed lagoons, and three-meter nautiloid predators that could propel their armored bodies in jet fashion, by squirting out water. The sun was 4 percent fainter than today, but carbon dioxide levels in the atmosphere were correspondingly higher, at times over five thousand parts per million (against four hundred today). The gradual strengthening of the sun since then has stayed approximately in balance with a general trend toward lower CO_2 levels, a balance that has kept global temperatures within a much narrower range than the one associated with the snowball earth. This is not sheer coincidence: living things have tended to draw down more CO_2 from the atmosphere in warmer conditions, thus making possible their own continued existence. But it has been only an approximate balance.

The first great check to Phanerozoic life took place some 443 million years ago. Gondwanaland, the largest continent by far throughout the early part of the eon, drifted down to the South Pole. That allowed an ice sheet, the first of the eon, to form on it. As ill luck had it, atmospheric carbon dioxide levels were under particular downward pressure at the same time, possibly for two main reasons. Firstly, the collision of the smaller continent Laurentia with an island chain had flung up a range of calcium-rich volcanic mountains, now part of the Appalachians. Those mountains accelerated a fundamental part of the carbon cycle: carbon weathering, whereby atmospheric carbon that has dissolved to form mildly acidic rainwater reacts with suitable rocks and then trickles into the sea to end up buried in the seafloor, often via a journey through the bodies of living things. Secondly, a shallow sea on the northern margin of Gondwanaland created an unusually large region of ideal conditions for marine life, and this had an effect similar to that of the accelerated carbon weathering. The marine animals used CO_2 from the atmosphere in making their shells, and the carbon was sequestered on the seabed when they died and sank.

The suggestion is that the conjunction of these three factors prompted a sharp glaciation, lasting half a million or a million years, which locked up enough water to make the sea level fall by as much as a hundred meters or more. It was no snowball earth, but a double pulse of extinction (once from the cooling, once from the returning warmth) wiped out a huge proportion— perhaps 86 percent—of all living species, and the ecosystems that grew back afterward were dramatically different from the ones that came before. This episode, known as the end-Ordovician mass extinction, nicely illustrates three things: the play of haphazard coincidence that has formed Phanerozoic history; the inseparability of organic and inorganic factors (the reason why Chakrabarty's distinction between living and nonliving things can only ever be relative, not absolute); and the centrality of climate to the course of the eon.

The end-Ordovician was the first of the five big mass extinctions that have punctuated the story of the Phanerozoic. The second, the Late Devonian extinction, was a long-drawn-out set of ecosystem collapses taking place some 70 to 84 million years later. In the interim Laurentia and two other small continents, Baltica and Avalonia, became sutured together, and the land filled with life. The continents had previously been places of bare rock and loose sediment, with some algae and fungi at their damp margins, because living on land means coping with desiccation, scorching, and the pull of gravity. In this period, however, marshy spore-bearing plants developed the technologies—seeds, rigid stems, broad leaves, and nutrient-carrying vascular tissue—to colonize the interior. By its end, tall forests of

Archaeopteris, a high-stemmed tree with a conical crown, covered the land. Their shade made terrestrial life possible for salamander-like amphibians the size of large pigs.

But the rise of *Archaeopteris* had ripple effects. Not only did the trees build themselves from atmospheric carbon, but just as importantly their transpiration intensified the circulation of water vapor and their roots broke up the deep rocky soil. Both of the latter influences accelerated the carbon weathering process, in which CO_2 reacts with rock and ends up locked away on the seabed. The result was a fall in atmospheric carbon dioxide, implying another phase of global cooling to end what had been a generally warm and dry interval. The forests may also have accelerated nutrient runoff from the continents, stimulating plankton growth that exhausted the oxygen in surface waters. The cooling climate and pulses of anoxia might have been key factors in the ruin of the communities gathered around the immense coral reefs of the tropics. The land plants suspected of having sparked the glaciations suffered as well. Over some millions of years, 75 percent of species, mostly marine, went extinct.

Afterward, life set out in a new direction once again. The 100 million years after the Devonian event saw the coming together of the supercontinent Pangaea. Gondwanaland met the Laurentian assemblage from the south, and Siberia closed in from the north to complete a thick C-shaped landmass that stretched almost from pole to pole. The grandest feature of Pangaea was a tall equatorial mountain range stretching across its center: the Hercynides, driven up by Gondwanaland's breaking on the Avalonian coast. Easterly trade winds traversed the surrounding ocean, Panthalassa, and rebounded from Pangaea's eastern shores in gigantic monsoon systems. In the time when Pangaea was forming, the monsoons fostered wetland forests that laid down the extensive coal deposits of the Carboniferous period. The supercontinent's inner zones, by contrast, were so far from the sea that they became dusty, salty deserts. The temperature difference between its

Map 1. Pangaea before the Great Dying, showing landmasses and the principal ocean currents.

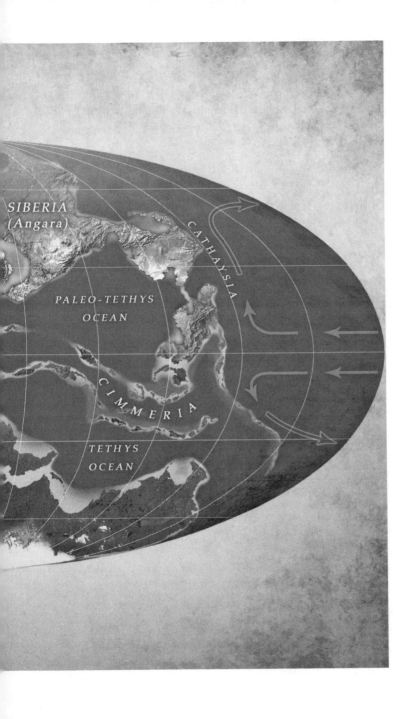

high and low latitudes was often very large. Scaly amphibians up to six meters long competed for the role of apex predator with similarly sized reptiles and warm-blooded therapsids, the ancestors of mammals. Millipedes grew larger than humans, and dragonflies' wingspans could reach seventy-five centimeters.

Highly oxygenated air (perhaps up to 35 percent, against 21 percent today) made it possible for such giant insects to breathe, and atmospheric carbon levels fell again—locked away in the coal—for a long stretch in the middle of the interval. That CO_2 decline produced a generally chillier world that saw the largest of all the Phanerozoic's glacial episodes. This did not mean another major extinction pulse, however, just a prolonged period in which the emergence of new species was much slowed down, before warmer, drier, and less oxygenated conditions returned. But a third great extinction did finally come, the first blow landing 260 million years ago, and the follow-up—the so-called Great Dying, the most dizzying horror yet faced by life on earth—8 million years later. Together, they destroyed 96 percent of all living species.

The magnitude of this third mass extinction, the Permian event, is reflected in the structure of the geological timescale. The Phanerozoic eon is split into three *eras*. Up to now we have been in the Paleozoic era, that of "old life." The Permian crisis is the dividing line between the 289 million years of the Paleozoic and the 186 million years of the Mesozoic ("middle life") that followed.

The sheer scale of the Permian disaster makes it hard to explain, but the key was probably a volcanic outpouring in Pangaea's northern province, Siberia, that covered an area twice the size of Mexico in a layer of basalt up to three kilometers thick. Two or three million cubic kilometers of lava emerged in little more than half a million years, emitting carbon dioxide, sulfates, and chlorine. The lava rose to the surface through beds full of coal and hydrocarbon gases, which it turned to smoke. This Siberian Ragnarok may in turn have raised temperatures enough to destabilize the measureless reservoirs of methane, formed by bacteria-like archaean microorganisms, that exist in a frozen, highly pressurized state beneath ocean floors; the methane would have bubbled to the surface and sent temperatures still higher. The accumulating gases apparently produced rain so acidic that it reduced plant life to decomposing heaps, letting soil erode down to the bedrock. Dead organic matter would have piled up around the whole supercontinent's coast, thereby contributing to the gravest crisis of all, as the oceans themselves grew acidic and deoxygenated. This last calamity was begun by the rising CO_2; it then fed on itself as ever more things died and rotted and as Panthalassa's waters stratified into lethally anoxic horizontal bands.

The bombed-out landscapes of Pangaea that were left after the Permian event were dominated for millions of years by a few diminutive species. In place of the highly regionalized flora and fauna that preceded the catastrophe, there was *Lystrosaurus*, a stocky, flat-faced burrowing reptile whose population exploded in the absence of predators and in the presence of plenty of ferns to feed on. A couple of thin-shelled bivalve mollusks similarly monopolized the oceans. Most telling of all is the return of the stromatolites and thrombolites. These ancient and simple bacterial organisms had dominated the biosphere a billion years earlier, covering sea margins in sticky, smelly mats, but they were not built to survive alongside complex life-forms. The geography of the supercontinent meant that the interior remained arid and desertified, seasonal temperature swings remained extreme, and fierce winds blew. Once conditions began to grow less bleak, important new groups of animals rapidly appeared: ichthyosaurs, turtles, lizards, and the earliest true mammals and dinosaurs. But thanks to intermittent further ripples of extinction, ecosystems may never have regained their old levels of complexity before this strange interval, the Triassic, came to an end.

Fifty million years after the Permian crisis there was yet another mass extinction, the death toll this time estimated at 80 percent of species.

Pangaea began to come apart along the old fault line between Gondwana-land and the rest, and lava welled up into the opening wound. That surge in volcanic activity meant that CO_2 levels tripled or more, temperatures leapt, and base layers of the food chain—seed plants on land (now overcooked) and shell-forming organisms at sea (unable to build shells properly in acidifying water)—disintegrated.

This Triassic event 201 million years ago, the fourth of the great extinctions, opened the way to the dinosaurs. They would dominate much of the world for the next 135 million years after diversifying rapidly into many ecological niches, including, of course, that of apex predator. *Tyrannosaurus* was the largest land carnivore ever to exist; the largest herbivorous dinosaurs were some ten times more massive. Coral reefs returned at the start of the age of dinosaurs. Cycads and conifers dominated the forests, although one of the interval's most significant developments, toward its end, was the coevolution of a rival cluster of species: the flowering plants and pollinating insects, originating in the tropics. Temperatures—and sea levels—were often high, especially around 100 million years ago, when the sea was warmer by six to seven degrees Celsius in the tropics than at present, and over ten degrees Celsius hotter at high latitudes. Gondwanaland and the northern landmasses pulled fully apart, leaving a channel between them—the Tethys Seaway—through which a powerful ocean current circled the earth from east to west. Gondwanaland itself then fragmented into the plainly recognizable shapes of South America, Africa, and India, as well as an Antarctica-Australia pairing; and by the end of the interval the continents were in a reasonable approximation of their current order.

A fifth great extinction ended the Mesozoic era and the time of the dinosaurs. It was the most spectacular one of all: a city-size extraterrestrial body—either an asteroid of dark rock or an icy, sparkling comet—collided with the earth at a speed of twenty or thirty kilometers a second. The impact, in shallow water close to where the Mexican town of Chicxulub now stands, was felt first in a shock wave of ferocious heat, a tsunami, and continent-scale wildfires lit by burning ejecta. There followed a months-long "impact winter" of blackout darkness and cold from the floating dust; acid rain; and some hundreds of thousands of years of global warming due to CO_2 released from the crater. The impact winter, killing off vegetation and marine photosynthesizers, was probably the biggest single contributor to the extinction of some 76 percent of species. Large, warm-blooded land animals suffered worst, and all nonavian dinosaurs died out abruptly; marine creatures that fed on detritus survived best.

FROM CHICXULUB TO PANAMA (66 MILLION TO 12,000 YEARS AGO)

The moment of the impact at Chicxulub is the beginning of the third and most recent era of the Phanerozoic eon: the Cenozoic (meaning "new life"). Each of these three Phanerozoic eras is further divided into *periods*. We have encountered the names of most of the periods already, because extinction events are named after the period that they bring to an end. The Paleozoic comprises the Cambrian, Ordovician, Silurian, Devonian, Carboniferous, and Permian periods. The Mesozoic is made up of the Triassic, the Jurassic, and the Cretaceous. And the Cenozoic, after some recent remodeling, is made up of the Paleogene ("old-born"), the Neogene ("newborn"), and the Quaternary (the "fourth": the remodeling has not been without some hitches). These twelve periods that make up the last 541 million years are themselves divided into a total of thirty-eight *epochs;* the Anthropocene would become the thirty-ninth. And the epochs in turn are each made up of between one and six *ages.* Thus the full range of time units marked on the International Chronostratigraphic Chart, from smallest to largest, is this: age, epoch, period, era, eon.

Fortunately, only a minority of these units are at all necessary for understanding the Anthropocene's context. The ages can safely be ignored for now as fine details of concern only to specialists, as can all the epochs older than the dinosaur extinction. Equally, the period divisions of the era *since* the dinosaur extinction can be passed over as clutter, and the difference between the Paleozoic and the Mesozoic can be guiltlessly forgotten. That leaves only two things. The first is the sequence of nine Phanerozoic periods from Cambrian to Cretaceous. This is the span of history from trilobites to *Tyrannosaurus*, from 541 to 66 million years ago. The second thing is the sequence of Cenozoic epochs covering the last 66 million years (a little less than one-eighth of the eon): seven epochs that may now be joined by an eighth. Their names all borrow from that of the era itself: they all end in -*cene*, "new." The nomenclature of the seven orthodox epochs follows a single trajectory that possesses a certain stiff, classical elegance. It begins with the "old new" of the Paleocene epoch, followed by the "dawn of the new," the Eocene, before it builds to a strangely hyperbolic conclusion in which the "most new" epoch called the Pleistocene is superseded by a "wholly new" or "entirely new" epoch, the Holocene. The full sequence is as follows: Paleocene, Eocene, Oligocene, Miocene, Pliocene, Pleistocene, and Holocene. Added on to this list, the Anthropocene does, admittedly, sound a little out of place.

The Paleocene, the first epoch of the Cenozoic era, began with a general spread of fern cover: opportunist species, recolonizing the wreckage left by the impact winter. But the distinctive feature of this epoch is the spread of colossal forests, dominated for the first time by flowering plants—a group that includes hardwoods like oak—rather than by conifers and their relatives. Temperatures were high, with no polar ice caps, and the vast expanses of woodland raised humidity worldwide. Some of the coelurosaur dinosaurs, albeit only the smaller members of that group rather than its large species like *Tyrannosaurus*, had survived the Chicxulub impact. They diversified fast in the Paleocene, producing species that ranged from shoreline waders to the South American forests' mightiest predators, the phorusrhacids. The phorusrhacids are nicknamed "terror birds": all the creatures that we now know as birds are these coelurosaurs, the surviving cousins of *Tyrannosaurus*.

Ten million years after Chicxulub, there came the first of the immediate precursors of the current moment: the first of six or seven transitions between one Cenozoic epoch and another. It was marked by a huge release of carbon into the atmosphere—totaling anything from three trillion metric tons to three times as much—over a period of some eight thousand to twenty thousand years. The source may have been a destabilization of methane clathrates as hypothesized for the Permian disaster (although not on the same scale), volcanic eruptions in the widening North Atlantic, or melting permafrost and drying peat in the Antarctic-Australian continent. In any case, the effect was a two-hundred-thousand-year burst of warmth, the Paleocene-Eocene Thermal Maximum, when the already high temperatures of the Paleocene rose some five to eight degrees Celsius higher. The heat was accompanied by more intense and seasonal rainstorms, and it brought on subtropical climates as far north as the Arctic circle. The Thermal Maximum is the largest climatic disturbance yet seen since the death of the land dinosaurs. It was a catastrophe for at least one key group of species, the sediment-dwelling foraminifera, or shell-forming marine microorganisms, who lost perhaps 50 percent of their diversity; but it was an opportunity for others. The continents that were by this point identifiable as Asia and North America were linked by a high-latitude land bridge, Beringia. The warm period coaxed many mammal species—ungulates and insectivorous primates among them—across the bridge from their Asian homelands. With this opening up of frontiers, the mammals were spreading into the niches left available by the absence of the dinosaurs.

The adventure of the Maximum marks the beginning of the era's second epoch, the Eocene. After this brief spike at its start, the early part of the Eocene saw the long-term temperature trend continue climbing to a peak level, fifty-two million years ago, that was around twelve degrees Celsius higher than modern times. From that point, temperatures started to decline again. The Indian tectonic plate—which had been heading northward ever since Gondwanaland fissured—began to make contact with Asia. The mammal clan branched out ever further. Tapirs emerged as the dominant herbivores, and within the course of one Eocene lineage a small, stream-dwelling, cloven-hoofed creature called *Indohyus* gave rise to a rodent-shaped terrestrial carnivore from which there descended an early whale, oceangoing and over five meters long.

The warm phase of the Cenozoic's first two epochs came to a definitive end thirty-four million years ago, with the transition from the Eocene epoch to the Oligocene. Australia had pulled away from Antarctica, which was now firmly back over the South Pole. The channel between them grew deep enough for the circumpolar current of the Southern Ocean to wrap round Antarctica like a snake, cutting off the land inside it from the warm currents spreading down from the equator. An ice sheet grew on Antarctica, and it has endured there ever since. It had worldwide consequences: the spread of the ice and a sharp decline in CO_2 levels mutually reinforced one another, and sea surface temperatures probably fell by at least five degrees Celsius at high latitudes. Tropical sea temperatures also fell, although less than at the poles. The great postdinosaur forests began to shrink as drier grasslands took their place. There was a major series of extinctions, especially notable in Europe. Before the Oligocene, Europe had been an island cluster isolated from the Asian landmass, but now sea levels fell as water was locked up in Antarctic glaciers, and the Eurasian continent stood clear of the waters. Asian mammals moved west and overwhelmed the often cold-blooded European fauna: the cooler, more arid world of the Oligocene saw rodents and the cat and dog families spread widely. In Asia again, this was the epoch of *Indricotherium transouralicum*, a sort of elongated rhinoceros that—at an average of eleven metric tons or perhaps more—rivals later mammoth species for the title of the largest ever land mammal.

The transition from the Oligocene to the eighteen million years of the fourth epoch, the Miocene, was less sharply marked than the preceding two. Against the grain of the cooling trend that began fifty-two million years ago, temperature and CO_2 levels held steady or even rose for the first half of the Miocene, before firmly restarting their downward journey midway through the epoch (fourteen million years ago). With cooler climates and drier summers, grasslands continued to expand, and as new species of grasses, weeds, and herbaceous plants multiplied and radiated, so too did the rat, mouse, and passerine songbird species that fed on their seeds, and the snakes that ate them in turn. Conversely, many herbivores went extinct near the end of the period, for the subtle reason that a change in the seasonality of rainfall (or, possibly, lower CO_2 levels) favored grasses of a more abrasive texture. Herbivores that did not happen to have long teeth had them worn away by eating, and starved.

These delicate evolutionary changes on the surface of the prairies ran alongside a wave of mountain formation. The African fragment of Gondwanaland had been slowly meeting with Eurasia; by now it was closing the equatorial Tethys Seaway that had opened as Pangaea fragmented. The faster-moving Indian plate, which had been diving under the Eurasian, began to crumple and buckle instead. Thus the Miocene saw the volcanic upthrust of the Himalayas, the easternmost part of the great Cenozoic belt of mountain chains that runs through the Hindu Kush, the Iranian and Anatolian plateaus, the Caucasus, and the Alps to the Atlas mountains in the west. The mountains were eroded even as they grew, accelerating the rock weathering process that strips CO_2 from the atmosphere. Near the end of the Miocene the only long east-west channel of the Tethys Seaway that still remained, the Mediterranean Sea, was almost completely shut off from the Atlantic at Gibraltar. Evaporation withered it away to a few dead hypersaline pools.

The end of the Miocene is another of geohistory's spectaculars. The Atlantic carved its way back into the Mediterranean and refilled it in a single gargantuan flood: 90 percent of the water volume seems to have returned in less—perhaps much less—than two years. The effects were global: the dry sea might have been sequestering 10 percent of all the salt in the world's oceans. This event, the Zanclean flood, was the fourth of the Cenozoic's epochal events; it marks the beginning of the Pliocene epoch, just over five million years ago. This is the epoch that featured in the Mauna Loa news reports as the last time when CO_2 levels were above four hundred parts per million. As the *New York Times'* snapshot of the epoch partly explains, and despite the long cooling trend that had preceded it, global temperatures for most of the Pliocene were still about three degrees Celsius warmer than in the early twenty-first century. Sea level was higher than today, ice sheets were mostly absent in the north, and the eastern Pacific was sweltering, perhaps owing to consistent El Niño conditions. Plant and animal life was by now evolutionarily similar to that of the present day, and the modern jigsaw of the continents was almost in place.

The one exception was the last and loneliest of the wandering post-Gondwanaland fragments, South America. While Africa and India were reforging their Pangaean bonds, the Atlantic continued to expand, and near the end of the Pliocene the two westering American continents finally hooked onto one another, not with a mountain-building collision but with a volcanic island chain that arced across their divide, until the gaps filled in with sediment layers and a single landmass took shape. The effects of the formation of the Isthmus of Panama were dramatic. Animals crossed it in either direction, but the dog, cat, deer, bear, camel, squirrel, and elephant families coming from the cosmopolitan north were far more successful than the endemic species of the south—from the "terror birds" and elephant-size ground sloths to anteaters, armadillos, and marsupial mammals—who underwent a shattering extinction.

The greatest consequence of the Pliocene-epoch formation of the Isthmus of Panama, however, was that it split the Atlantic from the Pacific. In doing so, it reorganized the ocean currents that underpin climate patterns, and another decisive coming-together of causal mechanisms took place to produce a sharp change of state in the earth system. Surface ocean currents normally flow from warm to cool areas—that is, from the tropics to the poles—sharing out heat as they go. The isthmus tended to deflect warm Atlantic water toward the Arctic, along with large aerial masses of water vapor. It also meant, however, that the Atlantic grew saltier as prevailing easterly winds drew freshwater from its surface and rained it into the Pacific. Saltier water sinks more readily, so the salinated surface waters of the North Atlantic no longer bore their equatorial heat all the way to the Arctic ocean, but sank and returned south around Iceland. This shortened heat transport helped the water vapor masses to fall as snow. The earth's gradually modulating orbit around the sun eventually provided a third conjoining factor: when the Arctic's summer sunshine levels crossed a tipping point, its annual snowmelt grew incomplete and glaciers formed. Once glaciers start to grow, positive feedback loops tend to keep them growing. Thus the Arctic froze, and the earth developed ice sheets at the North Pole to go with those in the south. Their formation marks the fifth epoch-level transition of the Cenozoic, and the beginning of the Pleistocene epoch, 2.58 million years ago.

With ice sheets now at both poles, the Pleistocene was utterly dominated by its recurring cycle of glacial periods (sometimes informally called ice ages, although that phrase can also refer to the epoch as a whole) separated by warmer interglacials. The strongest beat within these cycles came on a 41,000-year interval until about a million years ago. At that point, a slower 100,000-year dominant rhythm, with more intense freezes and thaws, took hold. The timing of the cycles was governed by small variations in the tilt, wobble, and distance of the earth relative to the sun, which affected the way in which the sun's heat was distributed over the planet's surface during the course of each year. Those changes were far too small in themselves to cause the glacial cycles, but they triggered elaborate feedback mechanisms, including changes in ocean circulation and the "biological pump" described above (whereby stronger winds carried nutrients into the oceans, and more carbon-bodied marine microorganisms were born, died, and were buried on the seafloor). Between them, those mechanisms repeatedly transformed the balance of the carbon cycle, and with it the state of the earth.

The temperature difference between Pleistocene glacials and interglacials was far greater at high latitudes than in the tropics, so the overall global temperature change is hard to calculate, but it is unlikely to have been less, or a great deal more, than five degrees Celsius. The pattern of the 100,000-year cycles involved a warm interglacial, with a climate comparable to the present day, lasting for a tenth of the cycle (sometimes rather less, occasionally much more), followed by a jolting slide into ever icier conditions that would reach a maximum shortly before an extremely rapid spell of warming brought the cycle back to the start. As each glacial period developed, polar ice sheets and mountain glaciers expanded, sea levels fell, rains dried up, and animals and plants converged in restricted habitats closer to the equator. Overlying the basic coldness and aridity, both seasonal variations in climate and climatic unpredictability from year to year were far more intense in glacial than interglacial periods. In the interglacials, flora and fauna were free to rush back up to higher latitudes: grasslands and cold-tolerant grazing animals like reindeer led the way, followed by birch and other pioneer trees, and then weightier slow-growing forest ecosystems.

The challenging conditions apparently did mean that overall biodiversity was reduced compared to earlier parts of the Cenozoic era, but not dramatically so. The repeated breakup of populations into isolated refugia during glacial periods may even have tended to increase the number of species. The distance from the present is short enough to mean that many of the creatures living in the Pleistocene were those familiar today, except that far more large land mammal and bird species were among them. That brings the story almost up to date.

The Pleistocene epoch was the shortest of the Cenozoic epochs thus far, at just under 2.6 million years. Likewise, the Cenozoic era as a whole, at 66 million years, is appreciably shorter than the other two eras (trilobites to Great Dying; Great Dying to dinosaur extinction) in the Phanerozoic. But what I want to stress is that so far the survey of geologic time in this chapter has not involved any really head-spinning chronological incongruities. At a length of 2.58 million years, the Pleistocene is only a small part of the entire Phanerozoic eon, but it is not an altogether minuscule part. Specifically, the Pleistocene occupies just under one two-hundredth of the whole Phanerozoic. That means that all the way from the snowball earth to the Pleistocene epoch, millions of years—rather than thousands or billions—is the appropriate unit in which to count. The whole eon up to the time of the glacial cycles can be imagined more or less well as a single story of world-ecological change.

This, then, is the deep-time context that gives meaning to the current transition from the Holocene epoch to the newborn Anthropocene. As I have said, there is certainly no need to attend to every detail of the story in order to make sense of the new epoch. But gazing back through geologic time is the only way in which to get a feel for the kind of change that is taking place in the birth of the Anthropocene. The great mass extinctions are the most colorful examples among many others of what an unpredictable and capricious course life on earth has taken. At arbitrary intervals an ill-starred configuration of continents, a self-sabotaging turn in land-plant evolution, bursts of volcanism, and an extraterrestrial impact have comprehensively transformed the stock of living things. The species that survived those extinctions did so not because of how well they had been adapted to the previous conditions but mainly by luck. In somewhat less eye-catching fashion, the whole story of the eon bears witness to the same laws of coincidence, happenstance, and casual destruction evident in the case of the herbivores of the Miocene epoch, among whom the species that survived were those that happened to have long enough teeth to cope with the abrasive new kinds of grass.

Above all, the geohistory of the Phanerozoic reveals no inevitable progress toward the coming of human beings and, for nearly all of its length, no hint that human intelligence would appear. The crucial message is that the birth of the Anthropocene is not an event that fulfills, transcends, and accelerates the dull, slow time of nature's rule; nor the start of a time belonging to humans that automatically stands in contrast to all merely preparatory nonhuman time; nor the moment when the earth finally completes the process of evolving a species that can breach its long-established limits and constraints. The idea of the Anthropocene simply couples the present crisis to the rest of geohistory, identifying it as yet another sharp and dangerous twist in the drama of deep time. The traces that humankind lays down for the far future will be one more unpredictable layer in the historical record, like the fossils of the *Archaeopteris* trees that monopolized the land surfaces 380 million years ago.

HUMAN BEINGS AND GEOLOGIC TIME

Nevertheless, some readers might by now be growing impatient. The question remains: when do *we* come into the picture? Never mind the Tethys Seaway and the evolution of flowering plants. For us, the really interesting issue is bound to be the arrival of modern, clever, cultured human beings, the *anthropos* of the Anthropocene. Where are we modern humans in relation to geologic time?

There is an extremely common answer to that question. It is often said that the time of modern humans occupies only "the blink of an eye in geologic time." Not only is that a terrible cliché, but it also has the potential to be deeply misleading. A better answer to the question would take issue with the very notion of modern as opposed to premodern human beings and deny that an absolute break between the nonhuman and the truly human can be found at any point in the evolution of life. (The notion of modern *societies* is a much more useful one, if still tricky to define.) The human species belongs within the multimillion-year course of geohistory as truly as every other species does. People of the present day are no more excluded from deep time, and no more superior to it, than *Indohyus* or *Indricotherium* were.

Writers who describe modern human existence as lasting for an eyeblink of geologic time usually envisage a blink lasting for about 50,000 years. The period around 80,000 to 40,000 years ago is the time of the supposed beginning of human "behavioral modernity." It is worth considering this idea of behavioral modernity in some detail, because if humankind was not cut loose from deep time when people first began to act in a modern way, it surely was not cut loose at any other point either. The tale of behavioral modernity has rather Eurocentric origins: it has traditionally been associated with the first *Homo sapiens* to make their way from Africa to Europe. It begins with an *H. sapiens* population in Africa undergoing a sharp cognitive and behavioral development. The development may have originated in a single decisive genetic mutation that rewired the brain, greatly improving the coordination between its various modules, or (less radically) in the crossing of a tipping point within a gradual process of mental development.[4] Either way, humans now became capable of far more complex symbolic social communication, a change recorded in their representational art, burial rituals, and long-distance trade but, at the time, no doubt centered in the elaboration of language.

The accomplishment of behavioral modernity sparked population growth and dispersal, with the first modern humans spreading beyond Africa or the western margins of Arabia some 70,000 to 60,000 years ago. Arriving in Europe, the moderns completely replaced the Neanderthals by 35,000 years ago; they had already reached Australia by 45,000 years ago and similarly replaced all earlier *Homo* species they encountered along the way. Their advanced, sociable brains meant they carried with them a pack of technologies: sophisticated stone blades, tools of bone and antler, hafted spears, red ochre pigment, and shells, teeth, and ivory to wear. By 30,000 years ago they had lined the walls of the Chauvet cave in southern France with inspired sketches of lions, mammoths, and rhinos. These dates are recent enough for further evolutionary "modernization" since then to have been confined to mere tweaks of genetic and cultural coevolution. (For example, the domestication of animals led to selection for the spread of genes that enabled people to continue to digest milk after infancy.) In other words, the first modern humans of 50,000 years ago are . . . us.

This story is still often told, and it rests on a perfectly legitimate archaeological basis. I have stressed how unpredictable and extravagant the course of geohistory has been throughout the Phanerozoic, and I do not want to take issue with the plausible idea that there were some correspondingly dramatic transformations in human behavior patterns between 80,000 and 40,000 years ago. But—in agreement with a brilliant polemic against the term by John J. Shea—I do want to argue that those events should not be understood as the coming of "behavioral modernity."[5] What Shea objects to is the framing of social change in the Later Stone Age as a once-and-for-all step forward into the definitive state of humankind. The idea of behavioral modernity rests on teleological thinking, as if "archaic" humans were simply an inferior or deficient version of the moderns, trying but failing to accomplish the advances that their successors finally achieved. It sets up a sharp dichotomy between the moderns and their unenlightened precursors, even as it downplays the differences between all those—from Chauvet cave-painters to Amazonian hunter-gatherers to Western city-dwellers—who are taken to share a common modernity. Modern behavior itself is reduced to a sometimes arbitrary-looking checklist of traits that are nevertheless taken to be self-evidently desirable and therefore irreversible.

This essentialist account of how *Homo sapiens* became fully or properly human is at odds with the spirit of the geohistorical narrative that I have been developing in this chapter. Moreover, it can easily go along with the idea that behavioral modernity marks an unprecedented break from gradual, Darwinian biological evolution, now comprehensively overtaken by self-accelerating, purposefully directed cultural evolution, from fire hearths to nuclear reactors. On the contrary, I have stressed that such breaks from Darwinian gradualism are far from unknown in earlier periods, when mass extinctions often prompted rapid, fortuitous evolutionary radiations. Once again: the emergence of the *anthropos* is a characteristic part of the turbulence of geohistory, not an alternative to it.

Abandoning the concept of behavioral modernity means abandoning the best hope that "we humans" might stand apart from earth history. What does human evolution look like in this alternative light? Many of the practices once known only from European sites and regarded as innovations of the last 50,000 years have now been found in much earlier, African contexts: red ochre was being ground into pigment well over 200,000 years ago, for instance.[6] Discoveries like these are crucial, but reconsidering behavioral modernity does not just mean pushing its start date back to the origins of *H. sapiens* perhaps 250,000–200,000 years ago. Not only is the origin of any species fuzzily defined in principle—a species can be distinguished only retrospectively and negatively, by way of its differences from the species to which it is related—but also the distinction between *H. sapiens* and closely related species, like the Neanderthals with whom "modern" humans interbred, is extremely unclear in practice. Nor is the solution to push back the beginning of humans' modernity still further, to the common ancestor of *H. sapiens* and the Neanderthals, a move that one might justify by noting that Neanderthals, like the moderns, buried their dead.

On the contrary, and as Shea argues, the most telling pieces of archaeological evidence are instead those that undermine the whole idea that particular technological changes were irreversible steps forward from less advanced to more advanced conditions. All such changes are better understood as local and specific responses to the fluctuating environments in which human groups operated. Southern African hunters used bows and arrows (normally considered a fairly advanced "modern" technology) 75,000–60,000 years ago, whereas their successors 60,000–50,000 years ago apparently did not. There is no evidence of a population crash that could have caused cultural "regression"; populations probably expanded at the time. Instead, changes in preferred food sources, hunting culture, or animals' behavior patterns—might hunters of the sixtieth millennium B.C. have switched to trapping and snaring their prey?—may have meant that people no longer considered arrow-making the best investment of their time and energy.[7] "Modern" behavior comes and goes in response to changing ecological conditions.

The longer history of the various hominin species is not yet at all well understood. Reconstructing it depends on finding reasonably intact hominin fossils, an exceptionally difficult task. Some landmark recent finds—the best known are those on the island of Flores and in the Denisova Cave, Siberia—and rapid, ongoing advances in paleogenetics mean that interpretations of hominin history are currently up in the air to a remarkable degree. The family tree of H. sapiens can be drawn up in any number of different ways, and the question of what drove hominin evolution invites limitless speculation: scholars disagree about the relative significance of bipedalism, tool use, language, delayed maturity, fire, drought, a lowered voice box, acute vision, weak vision, meat-eating, running, communal living, wide plains, steep valleys, the gendered division of labor, and much else besides.

That said, the first crucial event seems to have taken place late in the Miocene epoch, six or seven million years ago, at the time when abrasive grasses were driving herbivore species to extinction. In equatorial Africa, a single late Miocene population of tree-dwelling primates became divided into separate clusters, perhaps by some river or range of hills, and the groups' genetic structures drifted apart until they could no longer interbreed. One group became the present-day chimpanzees. The other—which seems mainly to have evolved further east, down the side of the continent from Eritrea to South Africa (although key fossils have puzzlingly been found in northern Chad)—tended to become less arboreal and more bipedal. This group diversified into an unknown number of species. Before three million years ago, in the Pliocene epoch, some of them had begun to make tools by battering lumps of basalt and other stones against anvil rocks, perhaps for use in butchery.[8] By around two million years ago, early in the Pleistocene, there lived a species capable of passing a test proposed by Daniel Lord Smail. Given clothes and haircuts, a *Homo ergaster* couple walking down a city street would probably prompt some odd looks rather than any calls to the police or to animal welfare officers.[9]

H. ergaster was apparently the first of several hominin species—again, the number is not known—to spread out of Africa into Eurasia. The latest of these were the *H. sapiens* migrants, from about seventy thousand years ago. They obliterated the Neanderthals and all other *Homo* species and subspecies in a process that appears to have involved a good deal of competition for resources, a little interbreeding (or, if you prefer, miscegenation), and perhaps some hunting (or, if you like, genocide).

The claim that the history of the human species is confined within the blink of an eye of geologic time is a boast disguised as self-deprecation. It shares with David Brower's sixth-day-of-creation sermon the belief that the slow time of nonhumans and the rush of human history are utterly incongruent and at odds with one another. The truth is that simply because they are an evolved species, humans have always belonged in deep time, even before the environmental crisis—the changes encapsulated in the news from Mauna Loa—turned that belonging into a politically urgent reality. The Anthropocene, in other words, does not fasten human beings into a geological chronology from which they had previously been separate. There never was any such separation, although it is true that the timescales of plate tectonics once had very little practical relevance to the timescales of political economy. What the Anthropocene provides is a way to conceptualize human societies' participation in deep time, a participation that had always existed, and that has taken on practical salience under the peculiar circumstances of the present day. The specific nature of those circumstances is this: a lineage beginning with a creature akin to the Miocene primate *Sahelanthropus tchadensis* has lately grown forceful enough to give its name to an epochal transformation of the world.

THE SEVENTH EPOCH

The artists of the Chauvet cave painted their masterpieces around 30,000 years ago, a date that falls well before the end of the Pleistocene epoch. Up to now I have been speaking rather freely about geological epochs, and in particular about the transitions between them, as if those categories are more or less uniform and well defined. But there is evidently no space in between Chauvet and the present for a seventh Cenozoic epoch the same size as the others. The chronological scale changes sharply at this point. We have been dealing with time units of a similar order of magnitude all the way from the snowball earth, 640 million years ago, down to *Homo ergaster* and the Pleistocene epoch. Now we need to zoom in a thousand times over, to the scale of millennia instead of millions of years. The seventh defined epoch since the dinosaurs, the Holocene, lasted from 11,700 years ago until, perhaps, two generations ago. Its comparative brevity makes the idea of an "epoch," in the context of relatively recent geological time, a profoundly ambivalent one. Fortunately, this double meaning of the word *epoch* is useful rather than debilitating for thinking about the Anthropocene.

In one way of looking at things, the Holocene is an epoch by courtesy only. We have seen that the Pleistocene was dominated by major cycles—probably numbering nearly fifty—within which glacial periods, which were extremely cold by Phanerozoic standards, alternated with shorter, more temperate interglacials. The Holocene is simply the name given to the most recent of those interglacial spells. That means it could reasonably be seen as just a small subsection of the Pleistocene and, in climatic terms, not even a particularly remarkable one: just the initial warm-up phase of the eleventh in a run of prolonged and intense glacial cycles that began a million years ago.[10]

Nonetheless, there are good practical reasons for crediting the Holocene with the status of an epoch. Geologic time divisions are not intended to be immutable natural facts about the history of the planet. Instead, all those divisions are justified primarily by their usefulness to earth historians. The geological timescale is an attempt to frame and organize deep time in the way that best enables scholars to understand it, and geologists to interpret what they come across in the field. And since the beginnings of the study of geohistory, it has seemed important to give headline billing to the time during which the youngest features of the earth's surface were laid down: the features that were formed, as we now realize, since the last retreat of the glaciers. The soils and sediments of the Holocene are prominent enough in the current makeup of the planet (and, as we will see, its fauna are distinctive enough) to justify calling it an epoch. That logic holds even though the Holocene is very different from any other epoch in the timescale. The stratigraphic profession is not like Colin Tudge or Matt Ridley in its attitude to the past and present. It has no pretensions to a perspective of Olympian impartiality, whereby the concerns of the world's present inhabitants are trivial in the grand scheme of things. On the contrary, the geological timescale is meant to serve the needs of the present, where the most recent interglacial period looms amply large enough to justify the status currently awarded to it.

The general acceptance of the Holocene as an geologic epoch is not just a terminological oddity, then. It is a demonstration that the meaning of the term *epoch* within geology is highly dependent on its context and potentially subject to dramatic variation. An epoch can resemble the Paleocene, Eocene, or Oligocene, all of which are time units of approximately the same kind, or it can be something like the Holocene, a unit barely one three-thousandth as long as the Eocene. That ambivalence matters very much in assessing the Anthropocene's potential status as an epoch.

The way Paul Crutzen tells the story, he first conceived of the Anthropocene precisely by way of contrast with the Holocene. It was casual references to the Holocene as the present epoch that made him blurt out: "We're not in the Holocene anymore. We're in the . . . the . . . the Anthropocene!" His argument for the Anthropocene arose directly from his conviction that the world had changed enough to pass outside the parameters within which it had remained for almost twelve thousand years. Scholars pursuing the line of enquiry that Crutzen sparked off have generally agreed with that view. In one sense, then, the Anthropocene takes its meaning primarily from the fact that it differs from the Holocene.

On the other hand, further reflection has made it increasingly clear that the world has changed, and is changing, far more than would be needed just to mark the end of the Holocene. The biosphere, especially, might be undergoing a reinvention comparable in its profundity to just a handful of other events in the history of life.[11] The current emergency is amply great enough to put an end not only to the Holocene but also to the larger Pleistocene-plus-Holocene state of affairs. Admittedly, current global warming is not necessarily on course to put a permanent end to the Pleistocene's sequence of glaciations. It is more likely to make the cycle skip one or more of its 100,000-year glacial episodes. But the transformations in the population and distribution of species; the accelerating number of outright extinctions; the altogether unprecedented speed of change in the carbon and nitrogen cycles; the acidification of the oceans; and the physical displacement of rocks, sediments, and minerals through building, mining, damming, and pollution: together, these constitute a general turn away from the earth system of the Pleistocene. Even in that unimaginable future time when the ice sheets of a new glacial period return, the makeup of the biosphere will be full of reminders of the present day. Or think again of an even more remote future, the time of Zalasiewicz's alien geologists. The swift warm-up that divides the Pleistocene from the Holocene would be invisible and insignificant to them, but not so, as we have seen, the changes that mark the coming of the Anthropocene.

We must look much further back in order to find any changes to the earth system that come close to the scale of the current phenomenon. Very possibly we must go at least as far back as the transition from the Pliocene into the Pleistocene 2.58 million years ago. That was the time when North and South America were joined together, the Atlantic and Pacific split apart, and the world began to swing in and out of glaciations during which up to a third of its land surface was thickly covered in ice. Revisions to the geological timescale should always be cautious and conservative, but present-day transformations seem to be fully sufficient to count as an epoch-level shift in any sense of the word.

Thanks to the double meaning of *epoch* in Quaternary science—it can refer to something like the Holocene, or to something like the Pleistocene—to speak of an Anthropocene epoch is to acknowledge both of these scales at once. The birth of the Anthropocene is at once the death of the Holocene and the death of the Pleistocene-plus-Holocene sequence. It concludes both a 12,000-year interglacial spell, and a larger 2.58-million-year spell in which repeated glacial cycles were overlain on a particular starting condition of the biosphere. The former is the more important in practical terms: the ending of the relatively stable Holocene interglacial is a deep challenge to civilization as such, as we will see in the next chapter. The latter is important because it lets us give current events their place within the whole history of complex life on earth, the Phanerozoic eon itself.

One might say that thinking of the Anthropocene as a new epoch in this ambivalent sense means that it is anchored to deep time at two different points, rather than just one. Or to use another metaphor, the ambivalence creates a way of looking at deep time with stereoscopic vision. In order to get a bearing on current environmental change in the context of deep time, it would be a good idea to commit two different dates to memory. The first of those dates is 2.58 million years ago: the beginning of the Pleistocene glacial cycles. The second date is 9700 B.C.: the beginning of the latest interglacial interval. It is the seventh "epoch" of the Cenozoic, the Holocene, that has created this useful equivocation in the very concept of an epoch. And it is to that seventh epoch, and its demise, that we must now turn.

CHAPTER FIVE

An Obituary for the Holocene

We have now come to grips with the deep history to which the idea of the Anthropocene gives access. To see yourself as living at the beginning of a new epoch of geologic time encourages an imaginative plunge through the 541 million years of the Phanerozoic eon. In Don McKay's words, it is an invitation to conceive of ordinary people of the present day as "members of deep time, along with trilobites and Ediacaran organisms": to "gain the gift of de-familiarization" by thinking of the distant past. That makes the times-cale of dinosaurs and supercontinents an indispensable context for under-standing the Anthropocene. A still more important context, however, is the most recent part of deep time. That is, the last phase of the Pleistocene epoch (marked by rapid climate changes and a wave of extinctions), together with the Holocene epoch, the interglacial interval that would go immediately below the Anthropocene in the geological timescale.

The birth of the Anthropocene is the death of the Holocene epoch, the epoch of "civilization." If you would like a single, simple reason for being concerned about the environmental crisis, this is it: human civilization has existed only in the Holocene so far, and no one knows what will happen to it once that setting is replaced by another. Most of this final chapter is taken up simply by a sketch of late Pleistocene and Holocene history: a description of the world that is ending. Recounting that story serves the fundamental purpose of this chapter, which is to examine how best to think about the relationship between the Holocene (including its prospective subdivision into three geological "ages") and the Anthropocene. It is essential to juxtapose the two in order to understand the present crisis—and, most importantly, to juxtapose them in a way that binds both epochs into the chaotic longer history of the planet.

THE CITIZENS OF THE HOLOCENE

The essential point has been made a thousand times. In the words of perhaps the most distinguished of all climate scientists, "global temperature has been relatively stable during the Holocene," which means that "civilization developed during a time of unusual climate stability."[1] James Hansen and Makiko Sato's observation opens up some pressing questions—what is *relative* or *unusual* stability, let alone *civilization?*—but its implications could hardly be greater. Why worry about human-caused global warming? Because it brings to an end a spell of nearly twelve thousand years during which the climate system was comparatively stable and temperate, and because human societies that are not based upon subsistence food gathering have hitherto existed only during that time of stability.

Global temperature has been "relatively stable" during the Holocene when compared to the Cenozoic era as a whole, the time since the death of the land dinosaurs, during which (as we have seen) global temperature fell by about 12°C over millions of years. It has also been relatively stable compared to most of the Pleistocene epoch, when the temperatures of large regions sometimes altered by 5°C or more within as little as a century. By contrast, the single biggest alteration in global temperature during the Holocene has been a long-term cooling of about 0.7°C in the five and a half millennia preceding the twentieth century. The biggest short-term change was a centuries-long cold episode centered 8200 years ago, which involved a cooling of 2°C in the North Atlantic, slighter cooling elsewhere in the Northern Hemisphere, and some warming in the South Atlantic, for an overall global temperature decline of a few tenths of a degree Celsius.[2] "Civilization" developed during this relatively stable period, in the sense that before 9700 B.C., the start of the Holocene epoch, no culture had developed writing or agriculture and nearly all human beings had always lived in simple bands as nomadic hunter-gatherers; only the most prosperous had formed some small sedentary villages.

Most of the world's governments have formally committed themselves to the goal of keeping global warming to a maximum of 2°C. Even that rise will take the planet outside the parameters of the Holocene climate system. The medium-term 4 to 6°C rise that seems increasingly likely would constitute a spectacular break with the world of the last 11,700 years. Once we have acknowledged the sufferings and disasters of the Holocene epoch, and its stereotypical social form whereby small elite classes parasitized peasant masses kept alive on bare-bones rations, it still remains true that democratic governments, scientific medicine, old-age pensions, mechanized transport, and dry well-insulated houses have so far existed only in a world in which global temperature varied by less than 1°C. The complex societies that made those things possible are entirely adapted to the earth system of the interglacial. But the Holocene epoch is at an end. It has reached its close even as the improved well-being that it offered to a few is finally, slowly becoming more accessible to a larger share of the world's population—a population that has itself grown a thousandfold since the start of the epoch. Perhaps its civilized achievements will, somehow, continue to flourish in a much hotter world. Perhaps they will not.

I have said that placing the environmental crisis in the context of deep time carries with it the risk of being seduced into an Olympian attitude whereby any merely human catastrophe is trivial in the grand scheme of things. The stratigraphic conception of the Anthropocene, however, encourages exactly the opposite state of mind: an emphatically time-bound perspective modeled on geologists' sense of the unique particularity of each turn of earth history. The way to achieve that concrete and specific sense of the present, and to escape the temptation to embrace airy Olympian futurology, is to pay as much attention to what is ending as to what is beginning. There is no real need to worry about what the planet will be like in that remote future when the Anthropocene comes to maturity. Instead, what matters about the Anthropocene, ethically and politically, is the way in which it is diverging from the Holocene, year by year and decade by decade. That means that our grasp of the Anthropocene can only be as good as our grasp of the Holocene. The birth of the Anthropocene is the death of the Holocene, and the problem of the twenty-first century is how to negotiate a way through the transition between these epochs. Which of the emerging kinds of divergence from the patterns established in the Holocene is it most important to avoid? Which can most bearably be tolerated? Which should we most desire to bring about?

An environmental politics for the years and decades ahead could start here: with an unhesitating attachment to the gifts of the Holocene—the hospitals, pianos, ballot boxes, and pizza joints that were unknown to the world before 9700 B.C.—combined with an urgent recognition of the need to respond in a radically new way to an earth system that has itself become something radically new. Attending to the Holocene is a way to articulate the risk that the complex web of civilization, with all its joys and its cruelties, is at risk of collapsing under its own weight. It means rejecting with equal force both primitivism and a business-as-usual attempt to make the status quo "sustainable." Instead, an ecological standpoint could be one of strategic affinity, sympathy, or mournful partisanship for the Holocene. It could be the point of view of people who feel themselves to be caught amid the labyrinthine fissures between one geological epoch and another. Environmentalists would be like patriots of a country that is being wiped from the map, obliged to look on their new homeland with the ironic distance and the keen sense of opportunity that are characteristic of exiles.

The Holocene deserves to be missed. But in proposing this quixotic loyalty to a dying epoch, I do not want to lose sight of two principles that I have invoked repeatedly throughout the previous chapters. The juxtaposition between the Holocene and the Anthropocene is of central importance; but, firstly, this dyad should not be thought of in isolation from the rest of deep time, and, secondly, we should not imagine that the Holocene was a time of stasis in the nonhuman world, a holiday from geohistory. Those two principles are closely connected. If the climate of the Holocene had been altogether unchanging, it would be like a firebreak of bare level ground separating present-day concerns from the climatic upheavals of deep time. But the reality is that both long-standing climate cycles and unpredictable climatic flukes continued to make their presence felt in the Holocene, just as they had in every preceding epoch. In short, we cannot rest content with James Hansen and Makiko Sato's (entirely accurate) observation that the Holocene has been a "relatively stable" interval of earth history. We must go on to think about what it means that the dying epoch has been only *relatively* stable. Failing to do that would mean neglecting, at the most crucial moment, the geohistorical principle that the earth has always been shaped by dynamic happenstance.

David Brower's fantasies about the "beautiful, organic wholeness" of the world before the Industrial Revolution never had a scientific basis. But until recently a similar idea on a smaller scale was part of the conventional wisdom in the earth sciences. It was thought that the oceans and atmosphere of the Holocene really were more or less unchanging. That view lasted until the 1990s, when a series of revelatory studies introduced a new paradigm. The current interglacial was shown to have been subject to a number of abrupt climate fluctuations, mostly stemming from the continuation of semiregular cycles that were already known from the Pleistocene epoch.[3] The fluctuations in the Holocene were on a much smaller scale than those of the preceding glacial period, which is why they were discovered later. However, even climate shifts that involved only modest movements in global average temperatures could be enough to change precipitation patterns across whole continents. Those sudden changes sometimes brought vulnerable civilizations crashing to the ground, as they did with the drought-stricken Egyptian Old Kingdom around 2200 B.C. and the Maya around A.D. 900. At other times they seem to have spurred innovations in political organization and the development of greater social complexity. The Holocene was not a blank canvas upon which human societies could express themselves. Its climate was not a dull, fixed reality, safely in the background and separate from human history. On the contrary, states and cultures have always been intimately shaped by complex, changeful feedback loops between politics and climate.

It is vital to keep this changeability of the Holocene climate in mind when trying to understand the Anthropocene. For one thing, it helps to make possible what I described in the previous chapter as a stereoscopic view of deep time. Understanding the Holocene does give us a long enough field of vision to appreciate the seriousness of the current predicament, but being able to understand the Holocene depends upon seeing how it fits inside the much larger Pleistocene, as the latest temperate spell within the long program of glacial cycles. And recognizing, as earth historians now do, that the long-term trends and cycles characteristic of the Pleistocene did not disappear in the most recent epoch, but continued in a moderated fashion, makes that stereoscopic vision much easier to achieve.

A second good reason for keeping in mind the fluctuations of the Holocene climate is that they make it easier to care about the fate of the geological interval that is now ending. If the nonhuman world had stayed stuck in place throughout the interglacial, except when it was altered by human actions, then it would be hard to take much interest in the survival of some-

An Obituary for the Holocene

thing so, well, boring. The dynamic progress of humankind would implicitly be set against its mere monotonous backdrop. As has become all too clear, dualism of that kind is a poor resource for ecological thinking and green politics. A nonhuman world conceived of as static and passive, helplessly in need of conservation and preservation, is bound to end up seeming like a tiresome burden, perhaps even to its most vocal defenders. Far better, surely, to attend to the tough and exciting reality of the Holocene's changefulness: its floods, its droughts, and its times and places of superabundant fertility. The geohistorical process whereby an intricate braid of human and nonhuman relationships tug back and forth has continued in full force throughout the Holocene epoch.

Hence the narrative that follows. This second story about deep time, now on the scale of thousands rather than millions of years, offers the exiled citizens of the Holocene a compressed and highly selective history of their disappearing homeland. The story illustrates how the Holocene can be perceived not abstractly (for instance, as the "emergence of civilization" against an inert backdrop) but as a living scene. That is, as something that one could take to heart, or for which one could feel sympathy or affinity. As with the story of the Phanerozoic eon in the previous chapter, the only way of even beginning to show what the death of the Holocene means in practice is by recounting something of what actually happened during the epoch. As with the Phanerozoic, again, I do not mean that memorizing the dates of the drying of the Sahara or the fall of the Akkadian empire is a prerequisite for making sense of the transition to the Anthropocene. I do mean, though, that sympathy for the Holocene relies in part on a working sense of the scale of historical change over the last twelve thousand years.

THE END OF THE PLEISTOCENE

The story of the Holocene epoch needs to begin with a prologue set in the interval that preceded it. We should start with the wave of migration out of

Africa that began around seventy thousand years ago—less out of partiality to human history than because of the effect that migration had on the non-human world.[4]

Even though the Holocene was a fairly typical subsection of the Pleistocene in climatic terms, its ecology was highly unusual. Specifically, it was the first interglacial after the extinction of most of the earth's large terrestrial animals. The first and worst extinction pulse was in Australia, seemingly concentrated more than forty thousand years ago. Of the continent's sixteen genera of large mammals, fourteen disappeared completely—among them marsupial lions, arboreal kangaroos, and *Diprotodon*, a monstrous combination of wombat and rhinoceros—along with all six genera of large reptiles and a giant meat-eating flightless bird. On the other continents, the extinctions ran from fifty thousand years ago down into the middle of the Holocene. Africa suffered least, although a horse and a giant buffalo were among its casualties. Eurasia's lost mammals include hippos and woolly rhinoceroses, giant deer, straight-tusked elephants, hyenas, cave lions, and Neanderthal humans. The Americas were the last continents to be hit, but about thirty-four genera of late Pleistocene North American mammals larger than forty-four kilograms, and fifty of their South American equivalents (just over 70 percent and 80 percent of the total, respectively) did not live to see the Holocene. They included mammoths and mastodons, gigantic ground sloths and saber-toothed cats, "the car-sized glyptodont and the three-toed litoptern, which resembled a horse with a camel's neck and a short elephantine trunk."[5]

Giant herbivores are powerful habitat engineers that break up stands of vegetation and accelerate nutrient cycles. Their widespread disappearance simplified ecosystems worldwide. In lowland temperate regions, mosaics of grassland, shrub, and woodland (attested to by pollen records) were replaced by uniform bands of closed forest, while the similarly productive dry, grassy "mammoth steppe" of higher latitudes in Asia and America gave way to mossy and waterlogged tundra. Coevolved plants that relied on the lost species for seed dispersal followed them into extinction.

The timing and selectivity of these extinctions—it helped to be small, nocturnal, or arboreal—make it all but certain that human hunting was the decisive factor. The most likely scenario is less a series of orgiastic bloodbaths than steady downward pressure on populations due to opportunistic killing by newly arrived hunting bands. Still, the effects might have been less drastic if ecosystems were not already under pressure: the extinctions took place as the Pleistocene climate cycle was going through its most hectic phase.

From 23,000 to 16,500 years ago, the period of glaciation that had begun about 115,000 years ago hit its coldest stage, the last glacial maximum (LGM). Twin ice sheets covered Canada and reached as far south as Indiana. Another sheet spread from Scandinavia to Wales and northern Germany. In the south, glaciers covered large parts of Patagonia and New Zealand, as well as some Australian and South African mountains. The temperature difference compared to the present was most acute at high latitudes. The extreme case was Greenland, which may have been twenty degrees Celsius colder than today. Western Europe was uninhabitable by humans above southwest France. In much of the world, however, aridity was just as significant as the cold (the two generally go together, because a fall in temperature reduces evaporation from the oceans). Africa was predominantly dry and deforested, and around four degrees Celsius cooler than at present; South America was mostly desert below the Tropic of Capricorn. The tropical western Pacific, oddly enough, may have been warmer than it is now. There was little permanent ice east of the Urals, but permafrost, and arid deserts with a wide penumbra of dry steppe and tundra, displaced forests and woodlands from most of Asia except for the southeast.[6]

The ice sheets drew sea levels down to 120 meters below their current height. In doing so they exposed tracts of land, nowadays submerged, that collectively were greater in extent than today's continental North America.

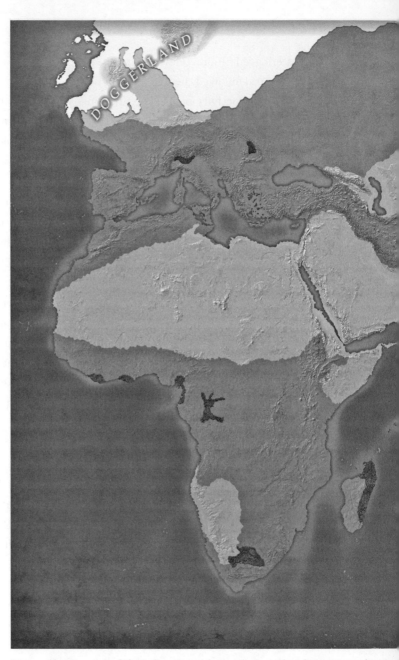

Map 2. Afro-Eurasia and the Indian Ocean during the Last Glacial Maximum, showing
Fennoscandian ice sheet, forest retreat and widespread deserts, and aquaterranes in th
terrestrial phase.

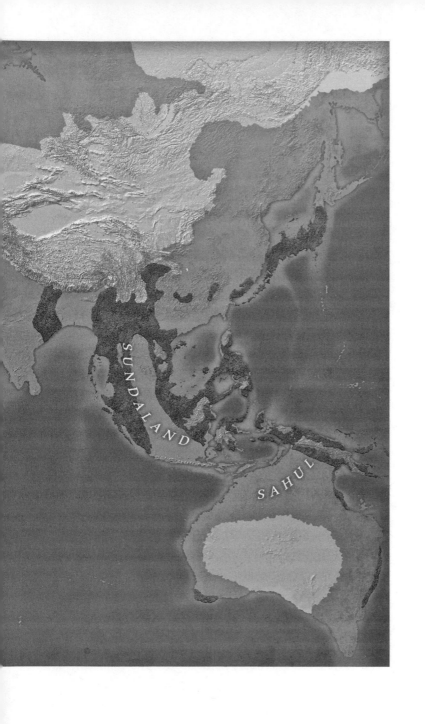

Those lands have been given the name "aquaterra": "a vast earth feature . . . which transforms back and forth among upland, wetland, and seafloor" as "the ocean gives and takes like a vast millennial tide."[7] Most of the aquaterra is at mid- and low latitudes, and in its terrestrial phase it forms coastlands, plains, and islands: it must often have been highly fertile, and must have played host to much of human history. It is almost entirely unexplored. The greatest of the lost aquaterranes was Sundaland, the remains of which make up the archipelago of Indonesia and Malaysia; it sustained what was then the world's largest tropical rain forest. A wide belt of land, Beringia, joined Alaska to Russia, just as Sri Lanka was linked to India and New Guinea to Australia. The Red Sea and the Sea of Japan were almost completely enclosed from the ocean. Not all of the low-lying land of the LGM is now submerged, however. In places the ground rose along with the sea, after having been relieved of a burden of ice.

Immediately after the LGM, temperatures—driven ultimately by changes in the earth's orbit, and proximately by meltwater pulses from the base of ice sheets, beginning in the far south—began to rise sharply. At this point the human population of Africa, Eurasia, and Australasia might have been about a million, and that of the Americas was probably zero. One scenario for the peopling of the New World has the western Canadian (Cordilleran) ice sheet retreating to expose the coastline soon after the LGM, and hunters from Beringia becoming the first humans in the Americas when they made their way south of the ice in boats, before dispersing—with admittedly extraordinary rapidity and lightness of archaeological footprint—widely enough to pop up at a woodland settlement in southern Chile 14,500–14,000 years ago. The exterminators of the mastodon would then be an offshoot of this migration or a later one, innovators who developed an improved stone-point hunting technology some 13,500 years ago.

Meanwhile, the late glacial climate changes were taking place. They were rapid and extreme across much of the world, and especially so in

An Obituary for the Holocene

Europe. We have seen that the Pleistocene glaciation as a whole was triggered partly by the Isthmus of Panama salinating the warm northward-flowing Atlantic current and thereby reducing heat transport to the Arctic. Glacial cycles within the Pleistocene were partly governed in their turn by the subsidiary waxing and waning of that shortened current. About 14,500 years ago, nudged by meltwater in the Southern Ocean, the North Atlantic current came out of a long quiescence and began funneling modern-day quantities of heat to Greenland and Europe. The temperatures recorded in Greenland ice sheets simply switched to the same levels found at the present, or even a little warmer. Then for 1500 years, Northern Hemisphere temperatures began to flicker uncertainly, with a general downward trend and a brief reprise of glacial conditions in Europe 14,000 years ago, but with enough warmth overall to drive a birch and poplar forest into Scandinavia on the heels of the retreating ice.

In the warm and wet conditions of the late glacial stage between 14,500 and 13,000 years ago, the hunter-gatherers of Southwest Asia—and specifically those bands who lived along the Mediterranean's eastern littoral—enjoyed perhaps the most favored conditions that any *Homo sapiens* group had yet experienced. It is worth lingering on these people, the Natufians, because the story of their successors and the successors of other nearby groups provides the main thread that we will trace through the history of the Holocene in this chapter. The Natufians gathered almonds, figs, raspberries, and pears, and caught wild boar, deer, and partridges. These varied foodstuffs supplemented the keystones of their economy: pistachios and acorns, migrating gazelles, and the stands of wild grasses—wheat and barley—that constituted the most nutritious group of plants then found anywhere in the world. They did not domesticate any plants or animals. That is, they did not produce any new strains or species, genetically distinct from wild populations, through selective harvesting and planting or selective breeding: they remained foragers, not farmers. But many Natufians adopted a settled rather than nomadic lifestyle, in villages built with wood and reeds on the ecotones between woodland and steppe, and they invested time and effort in sophisticated food-processing technology: sickles, grindstones, mortars, and storage silos. Moreover, their food waste suggests that they deliberately nurtured stands of edible plants—"wild gardens"—that grew near their villages.

The Natufians' prosperity lasted until the last paroxysm of the glaciation, a brutal cold snap known as the Younger Dryas. The partial melting of North America's gargantuan Laurentide ice sheet had left behind a correspondingly enormous pool of meltwater, known as Lake Agassiz, which was draining slowly southward into the Gulf of Mexico. Thirteen thousand years ago, melting glaciers opened up a new route northwest down the Mackenzie River, and Lake Agassiz sent a surge of cold freshwater mixed with icebergs out into the Arctic Ocean.[8] The climate system proved still to be on a post-glacial knife-edge. The surge shut down the North Atlantic warm current, and conditions much like those of the LGM—except that the ice sheets had by now shrunk considerably—returned across large parts of the Northern Hemisphere. Temperatures dropped in Greenland by ten degrees Celsius, in northwest Europe and eastern Canada by seven degrees Celsius in summer and even more in winter, and as far away as Chile by three degrees Celsius; even in China the coldness and desiccation was comparable to Europe's. In Southwest Asia, drought made the Natufians abandon their year-round villages. But the consequence there was not simply cultural regression. Whereas the sedentary villagers had been free simply to harvest local stands of wild grasses and then leave them to regrow, their nomadic successors apparently resorted to intensified management strategies, saving and carrying seeds in order to scatter and harvest them wherever they next set up camp for the summer. This still did not entail the production of genetically separate domestic species, but it meant that the Natufians' deliberate modifications of ecosystems continued to grow still more focused and profound.

The Younger Dryas lasted for some 1300 years. Then, 11,700 years ago, the North Atlantic warm current sprang back to life. Northern Hemisphere temperatures shot up by five degrees Celsius with breathtaking rapidity—the bulk of the rise was apparently completed in less than a decade—and remained there. The flickering changefulness of the late glacial climate was over, and the Holocene interglacial had begun. Compared to the 100,000-year glacial period as a whole, decade-to-decade climatic variability now dropped by a factor of five to ten, and this relative stability was to last throughout the Holocene. The stratotype, or golden spike, marking the base of the Holocene is found in a long, drilled cylinder of North Greenland ice stored at the University of Copenhagen. In three annual ice layers, most likely laid down at almost exactly 9700 B.C., the transformation of the climate is memorialized by a decrease in the proportion of deuterium relative to a heavy isotope of oxygen.[9]

Rain showered back down on Southwest Asia. In this steadier climate, the seed-planting techniques that had been supplementary to the collection of wild foods could increasingly become the basis of a radically new way of life in larger and more settled communities. It must seem as if the Natufians (although from the beginning of the Holocene they cease to be known by that name) were now better placed than ever before. Yet one of the most counterintuitive and best-established findings in archaeology is that the switch from foraging to farming does not improve living standards but degrades them. Agriculture is not invented as a means to improve the way people live. Instead, it evolves as a consequence of how populations interact with wild foodstuffs. And subsistence farmers before modern medicine were typically hungrier, more diseased, more disabled, less long-lived, and much harder-worked than their hunter-gatherer predecessors. Skeletal remains imply that between the late Pleistocene and 3000 B.C., the average height of adult males in the region from eastern Europe to North Africa fell by some ten to fifteen centimeters, indicating a calamitous decline in living standards.

Farming produces far more calories per acre than foraging, but populations soon grow even more rapidly than food resources. Women in early farming societies typically had two more babies each than their hunter-gatherer contemporaries, because sedentary families can have children at shorter intervals than nomads who must carry infants on their backs, and because new hands are always needed in the fields. Larger, more concentrated populations spread contagious diseases and diseases of filth; living in proximity to domestic animals allows other, still more virulent infections to cross the species barrier. The storage and distribution of food surpluses breeds both vermin and parasitical social elites. Diets dominated by a few staple carbohydrates are less healthy than hunter-gatherers' usually varied food intake and give rise to protein-deficiency and vitamin-deficiency diseases, from anemia to rickets. Labor time on food production and processing increases dramatically from the few hours a day typically employed by foragers. That often damages women's social standing most of all, as skilled resource-gathering gives way to a lifetime spent grinding grain and coping with the gross musculoskeletal damage that ensues. For both sexes a diet based on stone-pounded grains means eating grit with every meal and hence the agonizing decay of tooth enamel.[10] Nonetheless, once farming has first arisen it tends to spread, because even stunted and malnourished farmers soon become numerous enough to drive hunter-gatherers off the most desirable land.

Keeping this grim reality in mind has a particular usefulness when we come to think about the Anthropocene in light of the Holocene.[11] It provides a way to stop universalism sneaking in by the back door. I have argued that if the Anthropocene epoch is understood literally as a potential new stratigraphic interval in the geological timescale, then its proponents are not guilty of the self-interested bourgeois maneuver whereby the sectional interests of the rich are presented as being for the common good of all humankind (the undifferentiated *anthropos*). But seeing the birth of the

Anthropocene as the death of the Holocene might lead one to assume that "we" are, at least, all equally invested in the Holocene epoch. The idea might be: the whole world benefited from the stability of the last twelve thousand years, so we all have a shared interest, one that transcends politics, in keeping the world as close to Holocene norms as possible.

Recognizing that the climate of the Holocene was variable rather than a single perfect state of affairs, and that for some societies it proved extremely destructive, is one way of guarding against that sentimental and inaccurate line of thought. But a stronger guard is the realization that for the great majority of the Holocene, the living standards of the average human being did not show any signs of improvement—that in fact ordinary people's material prosperity often declined grotesquely with the switch from foraging to farming. Modern public-health medicine is probably the first thing to significantly reverse that trend. But it is still by no means clear that the standard of living of, say, a subsistence smallholder or an unemployed mother in Niger or Chad today is likely to be markedly better than the lifestyle of Natufian foragers before the end of the last glacial period. The Anthropocene is not a fall from Eden. The Holocene has by no means been a blessing to everyone, and it is certainly not something to be straightforwardly cherished; I have said that if one mourns its passing it should be in a critical, even ironic, frame of mind. And yet there is no denying that the agricultural transformation of the world is the fundamental prerequisite for democratic government, medical science, and all the rest of it.

TWELVE MILLENNIA

The idea of the Anthropocene creates an opportunity to look at the Holocene epoch as if from the outside. It means that we can try to see it as a whole for the first time, without foreshortening and at an approximately constant level of detail. Unlike the Phanerozoic eon, the Holocene is not readily split into chapters by well-established geologic time divisions (although plans are afoot in that regard, as we will see below). Instead, in the rest of this chapter I sketch its history by counting off one by one the twelve millennia of the epoch, like the twelve books of an epic poem.

In the interests of concision and memorability, and in a suitably experimental and playful spirit, I have given the millennia of the Holocene names or nicknames in what follows, as well as numbers. Most of those names are drawn from—and my narrative concentrates heavily upon—a single region of the world. The heart of that region is the so-called Fertile Crescent, which is in fact less a crescent than a bulbous arc that loops around the Syrian Desert and the northern part of the Arabian Desert in a series of low hills and floodplains. From west to east the arc consists of the Levant (homeland of the Natufians), the southeastern quarter of Anatolia, and Mesopotamia, the plains and deltas of the Tigris and Euphrates Rivers. Just to the east of the Crescent are the foothills of the Zagros Mountains on the edge of the Iranian Plateau; to the southwest are the Nile Valley and the Sahara; and to the west, the peninsulas and islands of the Aegean Sea.

This amorphous region was the homeland of the world's technological avant-garde for most of the epoch, which is part of the reason why its early history is much better known than that of anywhere else on earth. I have taken it as my focus mainly because its development is (uniquely, for now) well enough understood to allow for detailed discriminations between each millennium of the Holocene there. Another reason is that it can be thought of as the geographical hub or crossroads of the Old World. Included within it are parts of Africa, of Asia, and of Europe, the three continents that stretch away to its south, east, and northwest. By the fifteenth century A.D. and the beginning of the end-Holocene event, the region encircling the Natufians' homeland had been absorbed into a larger Indian Ocean trade world and surpassed in its population density and technological productivity by the agrarian empires of China and India. Nonetheless, it remained in that century one of Afro-Eurasia's archetypal contact zones. People, commodities, and ideas flowed in every direction through Baghdad and Aleppo on the Silk Route, through Cairo and Baghdad at the head of the Red Sea and Persian Gulf seaways, and across the Mediterranean through Constantinople/Istanbul (western Eurasia's largest city by far), while Byzantine, Timurid, Mamluk, Ottoman, and Safavid dynasties competed for imperial control. That was long after the end of the Younger Dryas, however.

+ + + +

The first millennium of the Holocene runs from 9700 B.C. to 8700 B.C. In the context of the Fertile Crescent and its environs, an appropriate name by which to remember it might be the Agrarian millennium. Even at this point there is no definite evidence of domestic crops in the region that were genetically and physically distinct from wild varieties: DNA probably continued to flow freely between managed plant populations and the wild. But the name suggests the way in which the cultivation of land in the Crescent—a practice perhaps somewhere between farming and gardening—continued to spread and intensify after the Younger Dryas. Wild foods like gazelles, boars, and nuts remained integral to people's diets. Now, though, they were eaten alongside einkorn wheat and emmer wheat, which were brought downriver from the hilly upper reaches of the Euphrates, sown in fields, defended from weeds, and supplied with water. Chickpeas spread from southern Turkey, and communal granaries were built to hold barley and oats.

In the Jordan Valley, a permanent settlement was established at Jericho, famously "the oldest city in the world," and grew to hold more than a thousand people as well as a squat, stone tower built eight meters high, probably used for ceremonial rather than defensive purposes. Jericho's satellite hamlet at Netiv Hagdud provides evidence of fields planted with lentils, and the satellite at Gilgal, of fig trees grown from cut branches that had been selected for their uncommonly sweet fruit. Villages were clusters of circular mud-brick dwellings, semisubterranean and thatched with reed, but the hilltop sacred center of Göbekli Tepe saw the construction of colossal limestone pillars in the shape of human torsos. Similarly shaped pillars were erected elsewhere in central and northern Mesopotamia, part of a ritual complex that also involved representations of snakes and birds as well as the skulls of humans and wild cattle, sometimes doused in animal blood.

The dawn of the Holocene saw ice and permafrost retreat from Europe, as hazel, birch, juniper, deer, and boars spread north (a temperature rise of seven degrees Celsius can expand a tree species' range a thousand kilometers toward the poles). Humans followed, hunting the forest mammals with bows and arrows, gathering hazelnuts, and catching eels in wicker traps. The richest center of this Mesolithic culture was probably Doggerland, a low-lying territory of marshes and narrow valleys on the northern edge of the continent. Wetlands expanded rapidly in the tropics, attested to by rising levels of atmospheric methane. Pottery, still unknown in the Fertile Crescent, was already a widespread and long-standing technology for cooking and display among the hunter-gatherers of China, Japan, and northeast Asia. The ceramics of the Jomon hunter-gatherers of Japan—whose diets involved acorns, fish, and smoked pork—grew increasingly sophisticated in this millennium as their first year-round villages emerged. Villages were established in North China as well; their inhabitants hunted deer, processed seeds, made tools of bone and antler, and may or may not have adopted sedentary lifestyles. Domesticated food plants, dissimilar to their wild progenitors, evolved in the Americas: nomadic cultivators in coastal Ecuador and the Colombian Andes domesticated squash and arrowroot, respectively.

The total human population might have been on the order of five million at the beginning of the millennium. Most human traces from the first several thousand years of the Holocene, however, are definitively lost. The societies of the Fertile Crescent had no doubt settled what is now the Persian Gulf, but rising sea levels must have drowned nearly all coastal villages of the Agrarian millennium long ago.

+ + + +

With the Fertile Crescent in mind, we could call the second millennium after the beginning of the Holocene (8700–7700 B.C.) the Pastoral millennium. Plant cultivation and animal herding evolved hand in hand in the Crescent, but the exploitation of animals (except for dogs, which had been widely domesticated as hunting companions long before) tended to run behind that of plants. Like cultivation, herding had deep roots. Some Pleistocene hunters tried to sustain their target populations by sparing females when they hunted males; the next steps were to lay out fodder when times were hard, and to capture and pen wild animals before they were wanted for slaughter. But in this millennium the first evidence appears at a series of sites—near Göbekli Tepe; further up the Euphrates; and to the west in central Anatolia—of true, lifelong herding of sheep and goats. Tamed pigs appear at one of the same sites. They may have arrived voluntarily, foraging for scraps. There are hints of incipient cattle domestication as well. Meanwhile, the domestication process was starting to be completed for cereal crops. Strains of emmer and einkorn wheat that were morphologically distinct and genetically isolated from their wild relatives appeared in the upper Euphrates (again, not far from Göbekli Tepe), while barley did so in the Jordan Valley and the Zagros.

The most vivid example of economic enterprise in the Pastoral millennium is the settlement of Cyprus. Colonists from the Crescent transported and released a whole package of species there, stocking their new island with einkorn, emmer, barley, flax, pigs, goats, cattle, deer, and even foxes. Social changes are attested to by the rectangular, two-story sandstone buildings constructed at Beidha in southern Jordan. Compared to the communal villages of the Agrarian millennium that merged with the surrounding fields, the thick walls of Beidha have been taken to suggest an increasing compartmentalization of social space, and the restriction of access to private dwellings.[12]

Lake Agassiz, the pool of Laurentide ice-sheet meltwater over North America, made its presence felt in another sharp flood. The North Atlantic warm current was weakened once again, and temperatures in the Atlantic dipped by two degrees Celsius around the middle of the millennium, bringing drought to Eurasia. Just afterward, however, the amount of summer sunlight striking the Northern Hemisphere reached its maximum level (some 8 percent higher than today), as the northern midsummer fell into sync with the moment when the earth was closest to the sun in its annual orbit. Because there is more land in the Northern Hemisphere than the Southern, that meant that the planet as a whole was approaching the warmest and wettest stage of the entire Holocene. The consequences were seen especially clearly in Africa. The Sahara was green: a vast savannah with a dense scattering of lakes that nourished gazelles, hippopotamuses, and elephants, sustained by monsoon rains drawn in from the Atlantic. In this millennium, if not before, the Sahara was repopulated by bands who hunted Barbary sheep, and who were ahead of the Southwest Asians both in the manufacture of pottery (probably used to boil grass seeds into porridge) and in their sophisticated management, which blurred into domestication, of cattle herds.

North America was also warming, but drying at the same time. That encouraged the spread of short-stemmed grasses, which in turn fostered herds of bison, a rare survivor of the late Pleistocene slaughter. Bison hunting was the basis of the Folsom culture centered on the Great Plains: a highly mobile society with low population density and minimalist settlements, best remembered for its sleek stone spear-points. And around the world, it's worth remembering, other foraging bands continued to pursue long-established ways of life. Thus in Spirit Cave in northern Thailand, rain forest

An Obituary for the Holocene

hunter-gatherers shared meals of nuts and seeds, fish and rats, and a menagerie of arboreal mammals (macaques, langurs, lorises, flying squirrels) that they may have brought down with projectiles poisoned with extracts of butternut and euphorbia.

+ + + +

One could think of the Holocene's third millennium (7700–6700 B.C.) as the Agricultural millennium. The innovations in human-nonhuman relationships that took place in the Fertile Crescent during the Agrarian and Pastoral millennia were now consolidated into a general package of ecological and social organization. The definitive domestications of goats, sheep, pigs, and cattle all seem to have shared the same pattern during the course of this millennium: final separation of the animals from wild populations in the center of the Crescent (or perhaps somewhat further east, for goats), followed by a spread down its arms west into the Levant and east into the Zagros. The animals changed physically, their horns and teeth diminishing in size. As the Crescent's agricultural package grew more uniform it also began to displace wild foods instead of supplementing them, both in the constitution of the typical diet and in the makeup of the landscape. Gazelle hunting had been the key to Southwest Asia's economy since before the LGM; but after millennia of overhunting, the most populous areas now experienced a rapid switch to domestic goats as the primary source of protein. Pioneer farming families and cultural transmission began to carry the Crescent's agricultural package far afield. Farming villages were established on fertile alluvial fans across the southern Iranian Plateau and all the way to its eastern edge, overlooking the plains of the Indus Valley. The agriculturalists brought cereals and goats with them, gathered dates, and would in time domesticate the local humped cattle, zebu, in place of the taurine cattle they had left behind.

The defining archaeological site of this period is Çatalhöyük, west of the Crescent proper in the southern Anatolian plain. It was a thirteen-hectare honeycomb of uniform, tightly packed houses, perhaps holding thirty-five hundred to eight thousand people. There were few streets—buildings were entered by trapdoors—and no public spaces, but an intense cultural life, seemingly preserved with few changes from one century to the next, was played out inside the houses. Almost every dwelling was dense with the skulls of humans and cattle, paintings of bulls and headless corpses, and burials underneath the plastered floors. The cultural complex earlier seen on a hilltop at Göbekli Tepe had apparently congealed into an intense ritualization of everyday life.[13]

CO_2 concentrations climbed to the 280 ppm level, at which they would remain until modern times, and global temperatures reached a plateau of warmth and humidity—the Holocene Climatic Optimum—that would broadly last until the seventh millennium. The eastward spread of the North American prairies attained its maximum extent. Around this time, the first sedentary villages within the Amazon rain forest were established. The villagers lived on fish and shellfish, arrowroot, fruit, and nuts. They piled discarded shells into bulky waste mounds, made large paintings on outcrops of rock, and reshaped the forests by cutting and perhaps planting trees, especially edible palms. Permanent villages also emerged in the distinct but interconnected population centers of North and South China. The southern sites, strung along the Yangtze River, left the clearest traces. Amid egalitarian circular dwellings wild rice was boiled in pottery vessels, perhaps having received some nurturing as it grew. Settled life was made possible mainly by a broad hunter-gatherer palette—water chestnut, soybeans, plums, peaches, ducks, water buffalo, muntjac deer—but pigs may also have been domesticated there at this early stage. Pots were buried alongside the dead.

+ + + +

The fourth millennium of the Holocene (6700–5700 B.C.) may be called the Diluvian, the millennium of the flood. The name comes from the phenom-

An Obituary for the Holocene

enon that paleoclimatologists call portentously the "8.2 kiloyear event." The Younger Dryas and Pastoral-millennium floods had not completed the dispersal of North America's ice and meltwater. A weakening remnant of the great Laurentide ice sheet still enclosed northern Canada, with a crescent of cold water lapping at its southern edge. A quarter of the way through the millennium, the lake water forced its way in between the ice and the land below it. Lake Agassiz-Ojibway drained definitively in a wide, high-pressure jet underneath the ice and out into Hudson's Bay, where icebergs carved deep parabolic scars into the seabed. The flood's volume has been estimated at 163,000 cubic kilometers, making it many times larger than the one that sparked the Younger Dryas.[14] It too caused a breakdown in the warm North Atlantic current, but with the planetary system as a whole now securely in its early Holocene warm phase, and because this time the freshwater was not injected into the sensitive heart of the Arctic Ocean, the ensuing climatic anomaly was much briefer and less severe than the Younger Dryas. Even so, it was the largest of the entire Holocene.

The anomaly, centered right in the middle of the millennium at eighty-two hundred years ago, was felt worldwide for as much as four centuries. Greenland temperatures fell by six degrees Celsius or more, Europe grew much wetter and cooler, and the Americas cooler and windier. The Asian and African monsoons were disastrously weakened, and the Fertile Crescent too was hit by drought. Sea levels, which had been rising throughout the Holocene, were all at once lifted another meter or two by the flood, enough to cause large-scale inundation of flat coastal lands.

A controversial theory holds that just after the end of the millennium the rising sea levels caused a miniature reprise of the Zanclean refilling of the Mediterranean, by making the Mediterranean itself catastrophically inundate the Black Sea, which in this theory had become a below-sea-level freshwater lake with a substantial (and ill-fated) human population around its banks. There is firmer evidence that the flood from Lake Agassiz

contributed to the calamity that hit the Jordan Valley. Ever since the foundation of Jericho, the valley had been a populous hotbed of agricultural innovation; plaster statues found there at 'Ain Ghazal may even reflect some of
the deepest roots of Babylonian religion. But in the second half of the Diluvian millennium its towns were abandoned, after a combination of deforestation, overgrazing, and the flood-born drought produced devastating soil
erosion.

Another long-standing center of culture faced an even swifter demise.
The flood accelerated the process whereby rising salt water gnawed away
Doggerland, the heart of the European hunter-gatherer world. The final
blow, the Storegga Slide, landed a few decades after the temperature-anom-
aly peak, seemingly in late autumn, when a section of coastal shelf off Nor-
way collapsed to the ocean floor. It produced a tsunami that struck the Shetland Islands as a wave twenty meters high before sweeping cataclysmically
through Doggerland's fertile estuaries and mudflats. By the end of the millennium Doggerland had been reduced to scattered islands and the sea had
divided Britain from continental Europe. New Guinea was likewise divided
from mainland Australia. Worldwide temperatures returned to their maximum level toward the close of the Diluvian millennium, although the Sahara
would never quite recover its previous monsoonal abundance.

+ + + +

The fifth (5700–4700 B.C.) and sixth (4700–3700 B.C.) millennia, the second
half of the warm and wet Holocene Climatic Optimum, are best thought of
as a pair. They could be called the Early Ubaid and the Late Ubaid. Together,
they might be imagined as constituting the classic period of the Holocene,
its early maturity. Glaciers retreated to their final Holocene boundaries, and
the 120-meter rise in sea level since the end of the LGM finally slowed and
then ceased at present-day heights. In the Fertile Crescent, a new district
came to the fore. Southern Mesopotamia—a sunbaked plain with little rain-

fall—was the last part of the Crescent to develop permanent settlements. But the Tigris and the Euphrates Rivers (especially the latter) had blanketed it with thick deposits of alluvial soil, and in the fifth millennium the construction of irrigation canals began to transform it into the site of a stupendously productive grain monoculture, generating stockpiles that would in time fuel the world's cutting edge of social and technological development. The Ubaid culture, distinguished by its building and especially pottery-making traditions, lasted on the alluvial plain almost throughout the two millennia, eventually spreading—seemingly by peaceful replication rather than conquest—into northern Mesopotamia and then as far afield as Anatolia and the Persian Gulf.

Ubaidian life was organized around a group of stable, conservative chiefdoms. The temple towns of Ur and Eridu, where offerings of fish were made to the water god Enki, each extracted cereal surpluses from perhaps a hundred square kilometers' worth of hinterland villages. Surpluses were redistributed in times of famine and traded for stones and timber from the northern hills. Otherwise, they sustained an elite class who exerted control over water, land, and labor, apparently without differentiating themselves from the masses by luxury consumption or expansionist militarism. The milking of sheep and goats, probably practiced on a small scale for millennia, appears to have yielded an increasingly important source of protein. The breeding of sheep with woolly coats allowed the emphasis in textile production to switch from flax to wool in about the sixth millennium of the Holocene, about the same time as the domestication of olives, grapes, and figs. This stability gave way only in the final centuries of the sixth millennium. The emergence of elaborate burials with copper trinkets, and of mass graves of young men killed in war, are the first signs of transition toward the cruel glories of the bronze-working states that would succeed the Ubaid culture.

Versions of the Fertile Crescent package diffused both southwestward and northwestward. In the Early Ubaid the raising of wheat, barley, sheep, and goats was adopted by the villagers of the Nile Valley and the cattle pas-toralists of the Sahara; a long-established agricultural toehold in the Aegean gave rise to a hybrid Mediterranean culture among peoples who combined wheat- and sheep-farming with fishing and foraging; and the Linearband-keramik farming culture emerged in the Hungarian plain, clearing wood-land from the fertile soils to make room for a utilitarian adaptation of the Crescent's agriculture based on cattle, grain, peas, timber longhouses, and simple incised pottery.

The spread of agriculture through Eurasia's western promontory is rela-tively well understood. In the first half of the fifth millennium the Linear-bandkeramik and its inheritors marched east and west across central Europe, while the Mediterranean hybrid culture diffused along the Atlantic seaboard. The agriculturalists were a mixture of colonizers from the south and the descendants of hunter-gatherers who switched allegiance to the new lifestyle; genetic evidence shows that the latter predominated. They traded with the nomadic hunting-gathering bands that had occupied Europe since the Younger Dryas. There are signs of social destabilization and increasingly chronic violence among the latter, but also of an "invisible bar-rier" some 150 kilometers inland from the northern coast, beyond which the Mesolithic coastal peoples—nomadic gatherers of fish, shellfish, seals, and waterfowl—resisted the spread of agriculturalism for more than a thousand years.[15] Only in the final third of the Late Ubaid millennium was Europe's Neolithic transformation completed, as the last maritime foragers of south-ern Scandinavia, Britain, and Ireland were pushed aside by farming societies.

Over the course of these millennia, the core Chinese agricultural pack-age fell into place: pigs and chickens, millet around the Yellow River to the north, and rice-paddy farming around the Yangtze to the south. The paint-ings and pottery of the Yangshao culture on the Yellow River bear witness to shamanistic ritual. On the Yangtze, multiroomed houses were built of clay and bamboo; women were buried with pieces of jade in their mouths. The first evidence of the autochthonous food-producing society that evolved in New Guinea dates to around the middle of the two millennia. In sharp contrast to the people of Southwest Asia—where agriculture meant clearing and irrigating fields to sow annual cereal grasses and herding livestock for protein and manure—New Guineans, who had no cereals or large animals to domesticate, specialized in vegeculture, the propagation of perennial plants. Since at least the early Holocene, New Guinean highlanders had broken up the rain forest with fire and transplanted yams, taro, and bananas into favored patches of soil. Now they began to intensify cultivation, clearing back the forest with ground stone adzes to lay out mounds of earth where they grew bananas, taro, and sugarcane. Four major domestications in the Andes likewise seem to date from this period: the potato, llama, alpaca, and guinea pig. The animals were probably brought in from the wild by semisedentary highland hunters who also cultivated quinoa, the full domestication of which may have come later.

Near the end of the Late Ubaid, Siberian hunters sailed to Wrangel Island, a patch of the drowned land of Beringia left stranded above the waterline. They found there, and hunted to extinction, the only population of mammoths that had survived into the Holocene, now dwarfed to human height.

+ + + +

The pace of both technological and climatic change accelerated dramatically in the seventh millennium (3700–2700 B.C.). It could be called the Hieroglyphic millennium: writing was among the period's many innovations. The changes in climate were ultimately driven by the changing geometry between the earth and sun. Ever since the Pastoral millennium, the amount of summer sunlight striking the Northern Hemisphere had slowly decreased. The decline in summer heating steadily weakened the force behind Northern Hemisphere monsoons, but the monsoons' feedback relationships with vegetation cover (among other things) meant that rainfall patterns often altered not gradually but in sudden, localized jumps. The seventh millennium saw a great cluster of these regional jumps to much drier conditions. In essence, it witnessed the emergence of the modern world's arid belt above and around the Tropic of Cancer, from the southwestern United States to the deserts of the Sahara, the Middle East, and Central Asia. It is tempting to link these climatic changes to social ones, and to see the struggle to cope with desiccation in the monsoon belt as the common factor behind the (very) approximately simultaneous emergence of complex societies on three continents.[16] Grand paradigms like that always work better as provocations than as dogma, but the widespread socioecological reorganizations of the Hieroglyphic millennium, like the temperature anomaly of the Diluvian millennium, undoubtedly constitute a major landmark for the view from the Anthropocene.

The most distinctive settlement of the Hieroglyphic millennium was Uruk, a day or two's walk north of Eridu and Ur. It already held ten thousand people by the start of the millennium. By its midpoint it was perhaps four times larger still. Uruk was a new kind of center, a novelty that might be captured by calling it the first city-state and the first colonial power. Its trading outposts, reached by pack-ass caravans and river rafts, were scattered from Iran to Anatolia; Uruk had commercial interests not only in timber but also in silver, gold, and lapis lazuli. The absence of equivalent Iranian or Anatolian outposts in Mesopotamia implies the beginning of core-and-periphery regional economic structures as opposed to networked ones. Uruk's art and high temples suggest that it was ruled not by mere chiefs but by kings with a mandate from the gods. Its prosperity was due to population density and craft specialization more than to technological innovation, which helps to explain the nature of its most memorable achievement: the art of writing had its banal origins in Uruk officials' search for a more efficient alternative to clay tablets in administrative stocktaking. Southern

Mesopotamian farmers had never relied on summer rainfall, so here aridification may have been more of a help than a hindrance, releasing waterlogged land for agriculture. Certainly, Uruk's collapse as a regional power in the second half of the millennium was followed by the rise of a congregation of wealthy city-states in its place. This was the hyperurban Sumerian polity, where, for a while near the end of the millennium, four-fifths of the population dwelt in towns that clustered around the holy city of Nippur.

Climate change affected the Sahara very differently. A sudden increase in windblown dust in Atlantic sediment records indicates that a widespread replacement of its monsoon-fed lakes and savannah by today's arid desert took place "extremely abruptly, within a few decades to centuries" around 3500 B.C.[17] The Sahara's cattle pastoralists were driven east into the Nile Valley. Soon afterward, hieroglyphic writing emerged there, perhaps inspired by rumors originating in Uruk, and the newly literate Naqada culture swept downstream to create the first unified Egyptian state. The coincidence is suggestive. Perhaps dynastic Egypt arose from cultural fusion between the cereal irrigation specialists of Naqada and the sophisticated cattle herders from the Sahara who had already built (at Nabta) the oldest known megalithic astronomical shrine. Or perhaps the Naqadans had become rich and imperialistic thanks to the exploitation of a new underclass of climate refugees.

Horses were domesticated for meat, milk, and transport by the nonagricultural Botai people on the cold steppes of northern Kazakhstan. Wheeled vehicles may have been invented in the same part of the world, although the technology spread so rapidly to Uruk and Europe that it is hard to tell; heavy wagons and carts drawn by oxen preceded horse chariots and fighting vehicles. The use of cattle in plowing—which opens up heavier soils to cultivation—was new or at least first widely adopted in the Hieroglyphic millennium. In the watershed of the Indus River a proto-urban culture grew up: the Kot Dijian, or Early Harappan, people built grid-plan mud-brick settlements; farmed barley, cattle, sheep, and goats; and adorned their polychrome ceramics with bull motifs. The cooling and drying trend in China coincided with the beginnings of Yangtze rice-farmers' spread toward Southeast Asia, and with the diffusion of an Austronesian-speaking rice-farming culture offshore to Taiwan: two demographic expansions that had much further to run.

+ + + +

Think of the eighth millennium (2700–1700 B.C.) as the Pharaonic: the Great Pyramid of Giza and the Sphinx were built in its first two centuries. The millennium approximates the lifespan of Egypt's Old and Middle Kingdoms, with their familiar pyramid tombs and stiffly academic court painting tradition. Contemporaneously with the Old Kingdom, urbanization reintensified in Mesopotamia. Timber, metals, and luxuries (ivory, pearls, lapis lazuli) were imported from across the surrounding region that I have described as the hub or crossroads of the Old World. Grain and woolen textiles were exported in return, while the kingdom of Akkad's precocious imperial conquests helped to underpin the system. The densest trade sinew ran through the Persian Gulf, plied by ships with reed-mat sails that linked Mesopotamia to a third economic core. This was the Indus Valley civilization of South Asia, which coalesced as a single polity across a huge floodplain region from the Ganges to the Iranian Plateau. At least five widely dispersed cities—Harappa and Mohenjo-daro the most famous—and innumerable smaller agricultural sites made up a single linguistic community. Its manufacturing districts absorbed raw materials from as far east as peninsular India and the Himalayan foothills; but despite the trade in prestige goods with Mesopotamia, Indus Valley social structures seem to have been relatively egalitarian.

Around 2200 B.C. there came a chaotic period of famine and cannibalism strongly associated with another major climate anomaly, the "4.2 kiloyear event." The Nile sunk calamitously low for a prolonged spell (at certain times and places one could easily walk across it), while Mesopotamia turned pro-

foundly cold, dry, and windy. The simultaneous desolation of the Old Kingdom and the Akkadian empire can reasonably be called civilization's first
dark age. The aridification affected low latitudes around the world: Amazonia suffered its most intense drought of the entire Holocene. There were
droughts and floods in China, acute aridity in North America, and colder,
wetter conditions in Europe. In Egypt and Mesopotamia, concentrated
urbanization was fully restored only after some centuries; Egypt's Middle
Kingdom was contemporary with the Babylonian empire of Hammurabi.

This was the first millennium of the trans-Eurasian exchange, of indirect
but life-changing contact between the distant societies of Eurasia. It was not
a question of colonization but of the sharing, appropriation, and emulation
of food cultures, which spread along mediating chains of farmers and Central Asian pastoralists. Many of the new food plants might first have been
adopted for ritual, ceremonial, or medicinal purposes. Chinese broomcorn
and foxtail millet, and possibly buckwheat, passed from eastern to western
Eurasia (that is, across a dividing line that could be drawn at the Pamir
Mountains). The classic Southwest Asian domesticates—wheat, barley, cattle, and sheep—crossed the other way, along with the craft of bronze metal-
lurgy. Foxtail millet, hemp, peaches, apricots, and East Asian rice reached
India. Indian sesame might have been transferred to Mesopotamia. Cattle
arrived on southern India's Deccan Plateau. Not having crops to fertilize at
first, the plateau's inhabitants burned the cattle dung in huge ceremonial
fires, marking their land with mounds of ash. The dung later served more
pragmatically to manure the local domesticates, mung bean and horse gram.
Rice-farming groups derived from those of the Yangtze came into contact
with the cultures of tropical Southeast Asia, right down to the Gulf of Thailand. Their rice- and pig-based lifeways hybridized with, rather than eras-
ing, the older ones based on foraging, fishing, and vegeculture.

Whereas the Hieroglyphic millennium had seen the first states develop in
Mesopotamia and Egypt, their South American equivalents first flourished

on the cusp of the seventh and eighth millennia. This was when the Norte Chico civilization arose in the meltwater-river valleys of the Andes, fueled by the unmatched marine resources—sardines and anchovies—of the upwelling Humboldt Current and influenced by another aspect of the general mid-Holocene climate reorganization, the beginning of the modern El Niño cycle. Populous, socially differentiated inland cities used the limited arable land largely to grow cotton for fishing nets, while coastal villages housed the fisherfolk. The cities, especially Caral, were dominated by spectacular stone pyramids rising over sunken plazas and were free of defensive structures: Norte Chico appears to have been a remarkably peaceful culture.

A major demographic transformation was brewing in West Africa. Around 2000 B.C., the rain forest and malarial wetlands in and around the Congo Basin contracted in a drying climate, and Bantu-speaking farmers of cattle, yams, and oil palms began to diffuse southward and eastward from their Cameroonian homeland. In the Great Lakes region of East Africa, meanwhile, a goat-herding pastoral culture had developed; bowls were made from pottery, lava, and pumice. The subtle, long-lasting changes that characterized Australia's various cultures are often hard to pin down to a particular millennium, but it may have been during the Pharaonic that dingoes were introduced from East Asia. They are implicated in the extinction of the thylacine and Tasmanian devil on the mainland. The millennium also saw a trend toward a cooler and drier climate in Australia, alongside the proliferation of lighter-weight stone tools that may reflect a growing population and intensified use of resources.

+ + + +

The ninth millennium of the Holocene (1700–700 B.C.)—we could call it the Assyrian—is like the fourth and the eighth in that it too is characterized by a disaster in the Afro-Eurasian hub region at its midpoint. Its first century or centuries also seem to have been a time of demographic contraction, and indeed one researcher has postulated a millennium-long "ecological depressive" downswing lasting for the whole period from 1700 B.C. to 700 B.C.[18] By now the world-changing bounty of southern Mesopotamia's fertile alluvial soils had become seriously depleted in the quest for agricultural surpluses. Intensive irrigation salinized the fields, so wheat crops had to be replaced by more salt-tolerant barley. Overgrazing (for wool) and upstream deforestation (for manufacturing fuel, shipbuilding, and housebuilding) caused erosion, siltation, and nutrient leaching. Grain yields diminished, and cities declined. Economic centrality shifted north to the steppe and foothills of the upper Tigris, where agriculture was based on rainfall rather than irrigation, and where a new cluster of large cities became the primary sink for the Fertile Crescent's ecological productivity. This was the homeland of the Middle Assyrian and Neo-Assyrian states, which persisted successively for the length of the millennium.

For all that, the first half of the millennium shows many continuities with the preceding one. After the early contraction, Egypt triumphantly reconsolidated itself under the storybook New Kingdom pharaohs: the queen Hatshepsut, the monotheist heretic Akhenaten, his consort Nefertiti and his son Tutankhamen, and Ramses the Great. Trade networks extended south toward the Horn of Africa for myrrh, ebony, and electrum, and north to draw in Baltic amber and English tin. The Aegean cultures—olives and grapes; blue sea and white marble; small boats trading oil lamps, saffron, and axe heads—flourished on Crete, the Cycladic islands, and the Peloponnese.

The disaster that struck in the middle of the millennium was swift. It has been interpreted in terms of the tight interdependence of the region's centralized, internationalist palace economies, such that an initial breakdown of trade routes began a disastrous chain reaction.[19] Drought and a swarm of earthquakes were among the triggers, although bands of migrant raiders—the mysterious Sea Peoples—and homegrown armies ultimately caused the most damage. Scores of cities from Greece to the Tigris were burned to the ground or abandoned within a generation around 1200 B.C.; only those of Egypt clung on, much impoverished. Again, the crisis was followed by a piecemeal rewiring of trade networks, notably those of the Phoenicians across the Mediterranean, and a rebuilding of cities, before the Assyrian reexpansion took command. Controversies about the historicity or otherwise of the biblical kingdom of David and Solomon are debates about what took place in the millennium's final third.

The first great Chinese states appeared in this millennium. On the Yangtze, a rich urbanized culture produced monumental bronze statues, but the millet-farming states of the North China plain that emerged from the Longshan culture—the Shang and Western Zhou dynasties, succeeding the half-legendary Xia—are better understood, having left behind extensive written records on bamboo slips and oracle bones. Shang social relations seem to have involved ancestor worship, kinship loyalties, and acute stratification of power, while the Western Zhou polity took a more rationalist and feudalist form. Ironworking is attested to in central Africa; it was probably invented there independently rather than spreading from the states to the north. Mesoamerica's distinctive cultural complex emerged among the Olmec chiefdoms and adjacent societies: earthen pyramids, fantastical sculptures including gigantic carved basalt heads, and symbolism concerned with water and with corn, the staple domesticate. The Austronesians who had sailed to Taiwan, and who then colonized parts of the Philippines and Indonesia, now became—along with their dogs, chickens, and pigs—the first mammals apart from bats to travel east of the Solomon Islands. Settlers raced from island to distant island in Melanesia, Micronesia, and Polynesia, as far east as Tonga and Samoa, drawn by untouched stockpiles of marine life and flightless birds.

+ + + +

The obvious name for the tenth millennium (700 B.C.–A.D. 300) is the Classical. On the north shore of the Mediterranean, the start of the millennium

An Obituary for the Holocene

almost coincides with the mythical date of the foundation of Rome, and with the earliest evidence of an urban settlement there. The *Iliad* and *Odyssey* were composed around the same time. The golden age of Periclean Athens took place in the millennium's third century; the Roman Republic gave way to the Empire in its seventh, just before the time of Jesus. In the Afro-Eurasian heartland itself, military, transport, political, and water-management technologies were making it increasingly possible to compile the region's diverse societies into cosmopolitan federations stretching between the Sahara, the grasslands of Central Asia, and the monsoonal world of South Asia. The successive dynastic polities that dominated the region may be seen as avatars of a single world empire, beginning with the Neo-Assyrian supremacy that broke down in the millennium's first century. It was followed by something even more expansive: the classical Persian empire of the Achaemenids. For two centuries a constellation of satrapies, some urbanized and others nomadic, were administered by a polyglot court based in the southwest of the Iranian Plateau.

A celebrated network of royal roads, traversed by relays of imperial messengers, held together the Pax Persica. After China initiated direct contact with the Persian world in the sixth century of the millennium, this network became the western portion of the silk roads. China had suffered ruinous instability for much of the first half of the period, but reunification under the despotic first Qin emperor (who was later buried with his terracotta army) had paved the way for the Confucianist and technocratic golden age of the Han dynasties. Traveling westward, Han China's envoys discovered a relatively loose-knit agglomeration of provincial kingdoms under the Parthian king of kings: by this time, the meteoric assault of Alexander the Great had triggered the breakup of the Achaemenid empire. The Fertile Crescent itself was split during this sixth century of the millennium between hostile Roman and Parthian imperia, reducing the Euphrates to a front line. More and more caravanserai headed west from China carrying silk and other textiles, pearls, and Indian steel; they returned with Spanish gold and silver, horses, perfumes, fruits, and wine. Philosophy traveled both ways. The first half of the millennium was Karl Jaspers's "Axial Age," the time of Confucius, Lao-tzu, the Buddha, Zoroaster (perhaps), Jeremiah, and Plato.

Early in the millennium, another round of desiccation opened pathways through the Central African rain forest along the coast and down the Sangha River. The Bantu farmers expanded south of the equator. After they reached the Great Lakes, a fusion culture developed—West African cattle and yams, East African domesticated sorghum and finger millet, and local ironworking techniques—one that toward the end of the period spread very rapidly south to the Cape, absorbing and displacing the hunter-gatherer societies in its path, and clearing large tracts of forest. At the same time, the Aksumite empire was drawing Africa's southeastern coast into the economic and eco-logical framework of the Afro-Eurasian crossroads. Aksum, ancient Ethio-pia, was the successor of an Arabian-influenced kingdom (D'mt) established early in the millennium, but it seems to have been more rooted than its pre-cursor in the long-standing urban cultures of the Upper Nile itself. The Aksumites controlled the maritime equivalent of the western silk roads—the Red Sea shipping between the Mediterranean and India—and they buried their kings beneath skyscraping granite stelae.

In Japan, the Jomon culture of sedentary hunter-gatherers had lasted throughout the Holocene, with a great flourishing in pottery sculpture in the seventh (or Hieroglyphic) millennium, when salmon runs and cultivated wild gardens supported a booming population. But around the middle of the Classical millennium, Yayoi farmers from Korea abruptly replaced the Jomon, arriving in the south and spreading to northern Honshu within two or three centuries. Principally rice farmers, the Yayoi also grew western Asian wheat and barley. The Adena and Hopewell cultures flourished in North America's Eastern Woodlands, the deciduous forest region between the Atlantic and the Plains. Sunflower, sumpweed, and goosefoot had been domesticated, and pottery was being manufactured, by the eighth (or Pharaonic) millennium, but the use of cultivated foodstuffs now increased rapidly and populations grew; rabbits and deer were hunted. Bands traded obsidian, copper, and shells over long distances, suggesting that the disparate cultures were largely peaceful albeit increasingly socially stratified. Log-lined collective graves and geometrical earthworks were built for the dead.

+ + + +

The eleventh millennium (A.D. 300–1300) probably cannot do without a compound name, unwieldy though it is to call it the Byzantine-Islamic millennium. Constantinople was founded as the second capital of the Roman world in A.D. 330. The Byzantine empire retained control of the Eastern Mediterranean littoral for several centuries, and the end of the millennium falls during its protracted final decline. For the latter two-thirds of the period, however, the hub of the Old World was incorporated within the Islamic civilization that blossomed out of Arabia. Muhammad preached during the fourth century of the millennium, and by the end of that century the pragmatic Umayyad dynasty in Damascus had already taken over from the Sasanian Persians as rulers of the core region from North Africa to the Iranian Plateau. The focus of philosophical and scientific brilliance turned to Baghdad under the Persianate Abbasid caliphate, before the final third of the millennium saw the Islamic world take the form of a looser congeries of independent states, now much more widespread than even the Achaemenid empire. Under its aegis, commodities, scholarship, and technologies intermixed from Iberia to Central Asia and India, and along Africa's frontage to the Indian Ocean, where the Swahili maritime civilization, linguistically Bantu and Islamic in faith, flourished in the second half of the millennium. On the Swahili coast Bantu, Arabic, and Persian merchants exchanged African gold, ambergris, ivory, and slaves for ceramics, ironware, and Indian textiles, forming the western flank of an integrated Indian Ocean-wide trade system.

To the east, China was fragmented again until the end of the millennium's third century. Before the middle of the period, however, a new cultural and commercial interstate complex had arisen, centered on a unified Tang-dynasty China and stretching deep into the dry Inner Asian grasslands. Silver, sugar, pepper and other spices, horses, and jade flowed inward from the south and from a circle of new-minted states—Japan, Korea, the Uighur empire, and Tibet—to which the Middle Kingdom exported its silks, ceramics, and metalwork. The seventh century of the millennium (and the ninth) saw the addition of a flourishing Khmer state based in Angkor, in the heart of the Southeast Asian tropics, with its temple-mountains and elaborate reservoir networks. By the millennium's eighth century, China's economic development had reached levels unprecedented anywhere in the world. Although smaller than its predecessor, the Song state was much richer and more urbanized than the Tang.

Austronesian adventurers hunted down the world's main remaining unpopulated landmasses with extraordinary brio, traversing the Pacific to settle Easter Island, Hawaii, and New Zealand (where they swiftly obliterated the moa, among many other birds), and crossing the Indian Ocean to reach Madagascar, with ill consequences for its pygmy hippos, giant terrestrial lemurs, and half-ton elephant birds. A climatic uptick in the North Atlantic during the final part of the millennium helped to foster the economies of the European peninsula, from whence a succession of marauders—the Crusaders—pillaged Byzantium and the Islamic west. A far greater upheaval followed. In the last century of the eleventh millennium, Eurasia underwent the explosive conquests of the Mongols. Once the rump of the Song state, melancholic and donnish, had fallen to Kublai Khan just before the millennium's end, the steppe nomads of Inner Asia were the overlords of four rapacious and cosmopolitan khanates that stretched from the Carpathian Mountains to the South China Sea and down into Mesopotamia.

The millennium corresponds just as well to the Classic and Postclassic periods in Mesoamerican history as it does to the Byzantine era. At the start of the millennium the cosmopolitan city of Teotihuacán was approaching the height of its glory, with grid-plan streets for its eighty thousand inhabitants surrounding a grandiose processional way lined with stepped pyramids. It collapsed around A.D. 550. Somewhat later, to the south of the Gulf of Mexico, the Mayan civilization reached its apex: a constellation of sophisticated, quarrelsome kingdoms that withered when an epic drought engendered environmental devastation, leaving its southern heartland permanently depopulated. Only the Maya's northern centerpiece, the pilgrimage center of Chichén Itzá, retained its splendor, nearly until the millennium's end. A characteristically Mesoamerican cultural complex spanned this wide geographical, temporal, and political variety: corn, beans, squash, and turkey; voluminous writing and especially calendar-making; feathers, gold, chocolate, and human sacrifice for the lords and the gods; ball courts for ritual game-playing. Mesoamerican agriculture spread north to underpin the Pueblo societies of southwest North America, at their height two-thirds of the way through the millennium with the mighty wooden houses or temples in the desert at Chaco Canyon. An almost uniquely sophisticated hunter-gatherer society flourished along the Pacific Northwest coast, where colossal salmon runs, whale hunting, and the produce of red cedar and hemlock forests sustained a steeply hierarchical society.

+ + + +

The best name for the twelfth and evidently final millennium of the Holocene (beginning in A.D. 1300) is the Columbian. Its first century brought the catastrophe of the Black Death. The horrors began in China and spread to the Mongols' Golden Horde, then on the tracks of war south into Persia and west to Europe. The first wave alone routinely killed a third of the population in the districts through which it passed. The Mongol khanates shattered; in China, a popular revolt produced the tightly centralized and increasingly dysfunctional Ming dynasty. The Ottomans, multiethnic provincial lords, seized the decaying Byzantine empire in the millennium's second century, making them the self-perceived heirs of the Roman caesars. They would go on to unite Constantinople both with Egypt and with what had once been the Fertile Crescent.[20] Trans-Saharan trade continued to sustain rich Islamic empires in West Africa, while the mercantile development of the Swahili coast began to be echoed in the far southern interior: the tall dry-stone palaces of the city-state of Great Zimbabwe reflected regional mastery over the supply of gold, ivory, copper, and salt to the seaports. In Mesoamerica, the Aztec states embarked on an expanding imperial project directed from the remarkable new city of Tenochtitlán, built on an artificial island in a lake and full of temples and rambling palaces. The Inca dynasty in the Andes was an even later upstart, with a still more splendid capital at Cuzco that oversaw an empire stretching five thousand kilometers from north to south, requiring much forced labor on road building.

But two centuries into the millennium, the set of changes that I have described as the end-Holocene event began to emerge. The European powers learned of the American continents and almost simultaneously the Portuguese established their route around the Cape of Good Hope into the Indian Ocean. Europe had always been out on the periphery of the Afro-Eurasian mercantile world, and it was still in recovery from a general crisis in its feudal economy that had been overdetermined by plague and climatic deterioration. But geographical expansion offered its ruling elites a way to reassert their economic standing through a new allegiance to the logic of perpetual capital accumulation, and their peninsular setting proved well placed for the exploitation of sea winds. Thus the evolution of capitalism spurred Europe's most developed territorial states, led by Portugal and Castile, in

An Obituary for the Holocene

concert with its most highly financialized entrepôts (Genoa, Venice, Florence, Antwerp), to an unremitting wave of overseas invasions.

There had been plenty of invaders, and many rich trade routes, before. But the divide between the Old and New Worlds had been constitutive for the Holocene, and few living things had crossed it since Beringia was submerged (a few seals, a few Vikings, some heroic migratory birds, the traders who sustained a very diffuse cultural complex across the Bering Strait). The socioecological fusion that the Europeans now oversaw was on a scale that differed from anything witnessed in the rest of the epoch. The human population replacement in the Americas, facilitated by smallpox, influenza, measles, whooping cough, and typhus, was by far the largest there had ever been. The Aztec empire fell to the conquistadores in 1521, and the Incas fell in 1533. The colossal bounty of the New World's resources slowly began to turn the balance of power in the Old World upside down. A febrile new episode of global ecological upheaval had begun, and it would continue to accelerate in the centuries that followed. This transformation in regimes of production, consumption, and energy flow constitutes the reshaping of the Holocene epoch into something else: the ongoing birth of the Anthropocene.

+ + + +

Writers on the present environmental crisis have often emphasized the comparative stability of the Holocene, describing it as a safe space within which civilization could flourish. With this survey I have tried to draw attention instead to its panoramic changeability, and to the unpredictable twists and turns that both humans and nonhumans have lived through since 9700 B.C. If the Holocene had always been the same, or if historical development during the epoch had always taken place along a single trajectory, it would not be so surprising that modern humans' grand ambitions and hard work had introduced a somewhat different note, or altered the planet's course a little.

What makes the fact of the Anthropocene remarkable is that the changes now under way are on another order of magnitude in relation to even the most spectacular ones of the preceding eleven thousand years. The climatic and ecological reorganizations of the Holocene, and its endlessly strange and disparate sights—mammoth hunters on a Siberian island, a lake-filled savannah hardening into the Sahara Desert, a village with trapdoors for streets and the bodies of ancestors laid beneath every dwelling, Korean colonizers conquering Japan with sacks of wheat from the far end of Asia—all go together on one side of the coin. Nuclear tests and the twentieth-century kaleidoscope of invasive species belong on the other.

THE AGES OF THE HOLOCENE

The Holocene epoch is represented on current editions of the International Chronostratigraphic Chart as a single block. But the introduction of the Anthropocene is not the only change under consideration for the topmost part of the geological timescale. The Holocene, as well as being brought to an end, might soon be formally subdivided. Outside the field of geology the latter plan has received far less attention than the former. Inside it, debates about these two potential revisions to the timescale have largely been kept separate from one another. Both tendencies seem regrettable. The documentation of the Anthropocene is best seen as only one part of a larger, ongoing reconsideration of how best to express the relationship between comparatively recent environmental history and the remote past of planetary time. The 1990s discovery of rapid and more or less worldwide climate change even within the Holocene—on a smaller scale, to be sure, than the fluctuations during the Pleistocene, but enough to have had profound effects on ecological and human communities—has helped to open up a new frontier for the craft of stratigraphic formalization. The changeability that scholars now recognize within the current interglacial interval might soon be reflected in the Chronostratigraphic Chart.

The golden spike that provides a definitive marker for the Pleistocene-Holocene boundary was ratified only in 2008. In 2012, an International Commission on Stratigraphy working group proposed splitting the Holocene epoch, for the first time, into three geological "ages."[21] The working group described an abundance of references to "Early," "Middle," and "Late" Holocene times in the existing scholarly literature, combined with wide variation from one paper to another in the way those terms were used, and noted the opportunities for inconvenience and confusion that result. They noted, too, scholars' otherwise increasing precision in the dating of Holocene events, and their growing success in correlating the timing of such events around the world. They argued that the time has come to resolve the entrenched but informal terminology for the phases of the Holocene into agreed-upon stratigraphic time divisions, in the interests of clearer and more precise communication between researchers.

The Holocene working group outlined a possible way forward. A formal boundary between the Early Holocene and the Middle Holocene will be fixed in the middle of the epoch's fourth millennium, the one I nicknamed the Diluvian. The dividing line will be the 8.2 kiloyear event, the furious drainage of Lake Agassiz-Ojibway into Hudson's Bay and the worldwide climatic anomaly that helped to bring down the societies of the Jordan Valley. The golden spike itself will be taken from the same Greenland ice core in which the GSSP for the Pleistocene-Holocene boundary is found. Two hundred sixty meters above the layers marking the birth of the Holocene, the ice reveals a powerful cooling signal (an alteration in the proportions of lighter and heavier oxygen isotopes) caused by the flood's interruption of the warm northward-flowing North Atlantic current. The second boundary, between the Middle Holocene and the Late Holocene, will be placed in the eighth, or Pharaonic, millennium. The Late Holocene will begin with the 4.2 kiloyear event, the breakdown in rainfall patterns across low- and midlatitude regions around 2200 B.C. that seemingly lay behind the near-simultaneous collapse of the Akkadian empire and the Egyptian Old Kingdom. The golden spike proposed for this boundary is a calcite stalagmite recovered from deep within a cave in the humid uplands of northeast India. There, another change in oxygen isotope ratios reflects the early stages of a centuries-long failure of the monsoon rains, a failure that was probably implicated in the breakdown of the Indus Valley civilization some time after the beginning of the dark ages in Mesopotamia and Egypt.

If the separate proposals made by the Holocene working group and the scholars of the Anthropocene were to be accepted by the geological profession, and if—for argument's sake—the golden spike adopted by the latter were to be based on the emergence of the global plutonium-239 fallout signal, we would be left with what seems at first like a strange way of packaging world history. The last completed interval represented on the International Chronostratigraphic Chart would be a geologic age called the Late Holocene. It would run from 2200 B.C. to A.D. 1952: from the fall of Egypt's Old Kingdom to the fall of its very last king, the British satrap Farouk. It would be younger, in other words, than cuneiform writing or the Sphinx. The transformative Hieroglyphic (seventh) millennium of the interglacial would be part of the preceding age, the Middle Holocene, which would date from the time when the North Sea swallowed Doggerland, the heartland of Europe's hunter-gatherers.

The two ICS working groups are by no means bedfellows or coconspirators. As it happens, the chair of the Holocene working group has been a thoughtful opponent of the stratigraphic Anthropocene. The project of subdividing the Holocene is more conventional and less controversial than the project of ratifying an Anthropocene epoch. (It causes less anxiety partly because the Holocene's prospective golden spikes are based on climate changes over which humans had no control.) That is not to say, however, that it is either uncontroversial or entirely conventional. The Holocene working group, like the working group on the Anthropocene, deals with timescales

An Obituary for the Holocene

far shorter than the millions of years involved in most geological research, and it seeks to assign dates for the changes that it scrutinizes with decadal precision, not far short of the annual resolution sought by scholars of the Anthropocene. Both groups are working close to the limits of stratigraphic science, in other words. Another point of connection is that interpretations of Holocene environments have long moved freely between considerations of human and nonhuman forcing agents. In studies of the climates, soils, and biodiversity of the last twelve thousand years, the lines between geology and archaeology are crossed routinely if they exist at all. In short, the subdivision of the Holocene is important to the idea of the Anthropocene because it provides a bridge between the Anthropocene and deep time. The idea of a geological epoch only sixty years old, and preeminently shaped by human hands, will seem less rebarbative if it is preceded by three geological ages each lasting around four thousand years and strongly marked by human presence.

The thermonuclear bombs of the Anthropocene; the monsoon rains whose vagaries would mark the beginning of the Late Holocene; the eccentric configuration of tectonic plates that made the planet a snowball in the Cryogenian period: all may soon be treated the same way in the stratigraphers' charts. A new system of periodization, running parallel to the old romantic and teleological system of the Stone, Bronze, and Iron Ages, but far more holistic in spirit, might soon have the weight of academic authority behind it. After the Tarantian, or Tarentian, age of the Pleistocene epoch (also lacking a GSSP as yet, but intended to comprise the last interglacial before the Holocene and the glacial period that followed) would come the Early, Middle, and Late ages of the Holocene epoch and then the newborn Anthropocene epoch.

A boundary line very near the top of the Chronostratigraphic Chart would mark the terminal crisis of the Late Holocene age and Holocene epoch. In retrospect, no doubt, we can see the seeds of the Anthropocene in the cultivated emmer wheat of what I called the Agrarian millennium. There has been no absolute divide between the Holocene and the Anthropocene; instead, there was a long process of intensifying pressures before a decisive transitional period was finally reached. But with that granted, we can recognize the beginning of the transitional period at the point when substantial biological exchange began between the world's two continental clusters. Geologically speaking, the globalizing order of the last five centuries can be thought of as the Holocene's terminal phase. Capitalist modernity is a strange-looking geological phenomenon, but it is nonetheless a real one: remember that life is just as much a geologic force as any other. The plutonium fallout spike of 1952–1980 can serve as a synecdoche for this crisis of the Holocene and the transition to the Anthropocene. The spike's inevitable survival in coral cores, ice sheets, and crater lake sediments provides a way of marking the transition to the thirty-ninth Phanerozoic epoch. It is the keystone of the new geological way of conceptualizing the recent span of earth history, the one in which *Homo sapiens* has played its vigorous part.

The new world of the Anthropocene is neither more nor less "natural" than the dying world of the Holocene. A disinterested observer who has no preferences as to whether hurricanes, algal blooms, or human beings are best able to flourish in the coming decades will see nothing to prefer in either one. From the constrained perspective of those loyal to the Holocene, however—those who feel an affinity for the achievements and hardships of the dying era—9700 B.C. is the single most significant point on the map of deep time. Everything that is happening for the first time since then is happening for the first time since the branches of fig trees were planted at Gilgal, and fields sown with lentils at Netiv Hagdud.

Conclusion

Not Even Past

The last thing that remains is to consider the implications of the stratigraphers' Anthropocene, and the theme of the passing of the Holocene, for contemporary environmental politics. It is vital to do so cautiously. It is not possible to deduce a detailed agenda for political action from the idea of the Anthropocene epoch—although I would like to think that it is possible to go a little way in that direction, and I will hazard some suggestions below. The Anthropocene provides a framework for understanding the modern ecological catastrophe, rather than a prescription for resolving it. It is a way of seeing, not a manifesto. And in particular, the hostile critics of the concept have overestimated the extent to which it entails specific policy enthusiasms—like geoengineering, rule by technocrats, or a preference for zoos over wilderness preservation—that are opposed by most of the Euro-American environmental movement. On the other hand, they have neglected its potential for intervening in the deep-down conceptual footings of environmentalism. As Don McKay put it, the new epoch offers a way for environmentally conscious citizens to see themselves as "members of deep time, along with trilobites and Ediacaran organisms," as "one expression of the ever-evolving planet." Its most direct political effect is to enfranchise such citizens who do not happen also to be specialists in paleontology, evolutionary biology, or the other sciences of deep time. By providing them with a standpoint from which to observe the winding course of earth history, the Anthropocene creates an opportunity to comprehend the environmental calamity in its full dimensions.

One major flaw in discussions of the Anthropocene so far has been the tendency to abandon the hard-won standpoint that arises from the stratigraphers' work as soon as it has been achieved. In other words, too many writers have been preoccupied with what comes *after* the Anthropocene as it is presently known. Taking seriously the geohistorical perspective encourages an altogether different attitude. It means accepting one's place in the actually existing moment of geologic time. The world is in the time of transition between the Holocene and the Anthropocene, and just confronting that transition is a large enough task for ecological activists.

Consider the well-known definition of the moral community that is often attributed to a Native American thinker or thinkers: the next seven generations. What I have called the end-Holocene event, the crisis of the old epoch intermingled with the birth pangs of the new, is set to continue in one way or another for the whole life of that community. (Who really thinks that global civilization will have reached a happy equilibrium point and settled down there for good in seven generations' time?) To recognize that is to see a new purpose for environmentalism. Not to escape from the crisis of the Holocene into a world made indefinitely "sustainable" and thereby liberated from geohistory, but something more demanding and more serious: to live within the crisis, and to struggle to influence its course by working for the survival of complex, pluralistic ecosystems. To adapt William Faulkner's phrase: the past epoch is not dead. It's not even past.

That said, the first political use that comes to mind for the Anthropocene is a much simpler one. It can work as a shock tactic. To say that the earth has changed so much that a whole new geological epoch has begun is a way of driving home the magnitude of recent damage to the living world. That role for the Anthropocene should not be dismissed lightly. There is no superfluity of scientifically grounded, single-word expressions with which to indicate the scale of the modern ecological emergency. The new term would be a worthwhile addition to the lexicon of the green movement even if that was its only purpose. Nonetheless, it has its problems when it is used in that way. The first is what might be called the David Brower syndrome: it runs the risk of making it sound as if change in the earth system is abnormal, thereby collapsing the history of the planet into a simple dichotomy between ancient natural stillness and modern anthropogenic depravity. The second problem comes from the name of the new epoch, which is what gives the concept of the Anthropocene its provocative force, but which risks giving the impression both that "we humans" are all equally at fault for environmental destruction, and that we now have the world under our collective thumb, to control it and direct it as we please. The belief that the concept of the Anthropocene is necessarily complicit in such retrograde worldviews is mistaken, however. And there is nothing wrong with using the word in a polemical fashion, to dramatize the urgency of the threats that confront the planet's living systems.

At the same time, there is evidently much more to the concept than just a way of grabbing public attention. The researchers who make up the Anthropocene Working Group have often said that assessing the stratigraphic basis for the Anthropocene should lead to a better understanding of environmental change. It can do so by setting the present crisis in the context of geological time, as I sought to do in chapters 4 and 5 of this book. Of all the landmarks on the map of deep time, the one with the greatest contemporary resonance is 9700 B.C., the end of the last glacial period and the beginning of the Holocene epoch. Since then, global temperature has stayed more or less within a one-degree-Celsius range, the human population has multiplied a thousandfold, and alternatives to the life of subsistence hunter-gatherers have emerged for the first time. If those agricultural alternatives often had many more victims than beneficiaries, they have nonetheless slowly yielded the prospect of emancipation for the masses of the world. The "unusual climate stability" that James Hansen and Makiko Sato described has been the precondition of the whole gamut of civilization.

The birth of the Anthropocene means that the planet is currently departing its twelve-thousand-year period of relative ecological dependability. That dependability was only relative, not absolute, and the geological record of earlier epochs shows that the planet has strong impulses toward sudden systemic change. For those reasons, the Holocene-Anthropocene dyad cannot be understood without locating those recent epochs within the long narrative of the Phanerozoic eon, or without recognizing the Holocene's own complex changeability. But given those provisos, it can hardly be said too often: the best reason to be worried about the current state of the planet is that the ecological setup characteristic of the entire epoch of complex societies is coming to an end.

Looked at from a stratigraphic point of view, the time of capitalist globalization since the fifteenth century can be understood as the crossing between two geological epochs. Taking a geological perspective on global capitalism might sound like an unlikely thing to do were it not that carbon dioxide levels have just risen to their highest level in three million years: thanks to disturbances like that, unprecedented for millions or many thousands of years, deep time has become politicized. To date the beginning of the epochal transition to the fifteenth century is not the same as saying that the rise of capitalism made the end of the Holocene inevitable. Far from it: it has taken a whole series of chance conjunctures to bring about such a great upheaval in the earth system. The close of the Holocene might be said to have begun with the linking of Afro-Eurasia to the Americas, an enterprise that happened to be overseen by emergent capitalist states that were simultaneously joining up the Atlantic and Indian oceans. The epoch's disintegration moved to another level thanks to the collision of demography, wage structures, coal technology, state power, consumer habits, and economic inputs from New World slave agriculture that sparked the Industrial Revolution. A third main stage involved the coming together of U.S. hegemony, Green Revolution agriculture, Fordist production and consumption, cheap oil, and a population boom, together making up the "Great Acceleration."

This multistage transition between epochs needs an emblem, a marker that can be picked out to represent the whole complex process. The worldwide plutonium isotope anomaly of 1952–80 is a strong candidate for that job, so it makes sense to fix the base of the Anthropocene at 1952, the year of Nasser's Egyptian coup, the Mau Mau uprising, and the Ivy Mike thermonuclear test shot. That choice of a recent golden spike should not be seen as downplaying the importance of the earlier stages of the end-Holocene crisis. Instead, it puts the emphasis on how many different forces had to intersect before together they could bring about a million-year change in the planetary system, a change recorded in a worldwide accumulation of layers of sediment that are perceptibly different from those of the Holocene and Pleistocene.

A formal dividing line placed as recently as 1952 also helps to underline the fact that the broader transition between epochs is by no means over and done. The end-Holocene event is still under way. The least satisfactory aspect of the Anthropocene debate so far has been its discussion of the relatively near future. That discussion has been dominated by a jejune back-and-forth about whether it is possible that there will be a "good Anthropocene," and by an overwhelming desire to predict what will succeed the current version of the Anthropocene. A final late phase of the epoch, in which society adopts a fresh set of values and in which natural resources are used sustainably? A new epoch of runaway climate change in which human population levels crash? A new epoch (the "Sustainocene"?) in which geoengineering and transnational democracy permanently stabilize the earth system? The modish "post-Anthropocene"? For the most part, this futurology lacks any grounding in a systematic theory of social change.[1] But its weakness only reflects in an intensified form a problem familiar to many traditions in environmental thought: an inability to explain convincingly how to get from here to there, from chronic environmental degradation to a sustainable world.

Much of the problem lies in the desideratum of "sustainability" itself, upon which a great deal of contemporary environmentalism, especially in its more institutional forms, depends. The apparently less conciliatory agenda based on the thesis of ecological "limits to growth"—lately back in fashion with a new lick of paint as a policy framework concerned with "planetary boundaries," and warmly adopted by several prominent theorists of the Anthropocene—has similar downsides.

The well-established radical argument against making sustainability an ultimate goal holds that the concept does not put forward any positive democratic agenda, but only advocates wishfully for the principle that the world's natural plenitude should be kept constant over time. At best, that principle is so ill-defined that it can easily be co-opted by mainstream capitalism. Business as usual can be passed off as "sustainable development," while green politics is reduced to an anemic clinging to the past, reliant on a pastoral fantasy about the perpetual harmony of unspoiled nature. At worst, it does not even need to be co-opted, since the goal of sustainability is already an explicit defense of the status quo: a managerial, efficiency-seeking principle with the avowed aim of securing the flow of natural resources required for the continued accumulation of capital. This critique is not always fair (*sustainability* does not express a single way of thinking any more than *the Anthropocene* does); but more often than not, sustainability is indeed recognizable as just a sanded-down version of another principle: the principle that there exist irremovable "limits to growth."

For theorists such as the doyen of bourgeois "ecologism," Andrew Dobson, the limits-to-growth thesis constitutes the "fundamental framework" that must unite the greens and distinguish them from the nonenvironmentalist mainstream. The modern version of this thesis crystallized with the collapse of the Bretton Woods monetary system in 1967-74: it was part of the way in which that collapse came to be framed as an "energy crisis" or "oil crisis."[2] The limits-to-growth principle holds that green politics must be built upon respect for the planet's carrying capacity, and must give priority to avoiding the exhaustion of natural systems and buffers. But limits to growth are *produced* through economic and ecological relations—via decisions about how resources are used, and what kinds of trade-offs can be made when extracting value from them—rather than imposed on the economy from the outside by the predefined material realities of the planet. That does not make it any less true that infinite consumption growth is impossible on a finite planet. It does mean, however, that invoking biophysical limits to growth cannot substitute for analysis of the relations that establish those limits in practice.

Environmentalists who promote sustainability and limits-to-growth environmentalism claim that their doctrines are set apart from nonenvironmentalist "gray" politics by the fact that they pay attention to the spatial constraints on human flourishing. For them, the good life depends upon reconciliation with the finite physical capacity of the nonhuman world. But the way in which they come to terms with that spatial constraint is by imagining the supersession of temporal constraints. If we take care not to run up against the limits to growth, our sustainable way of life may carry on indefinitely, or so it seems. So to talk of sustainability and steady-state economics is to deal in abstractions that would be equally applicable at any time; it is to engage in a romance of stasis.

The politics of the Anthropocene offers something different. The Anthropocene provides an alternative point of origin for green politics, a conceptual architecture with which to contest the logic of sustainability. A political standpoint designed for the birth of the Anthropocene should be one that begins with the full *spatiotemporal* reality of the ecosystems around it. It should be an approach to politics that arises out of the specific contradictions of the present ecological crisis, the transition between the Holocene and the Anthropocene. That is why it is such a pity that the debate about the Anthropocene keeps racing ahead to worry about what comes next: to the sustainable or apocalyptic future that is supposed to be in store now that the start of the Anthropocene has been and gone. Better to stay with the crisis. To inhabit it and seek to mold it. If we take geological timescales seriously, then the "post-Anthropocene" might as well be as far off as the heat death of the universe. We can scarcely even envisage a time when the Anthropocene will have separated itself from the last traces of the Holocene. For those reasons, ecological politics can be the politics of the end-Holocene event.

If environmentalists come to see themselves as working to shape the generations-long transition between two epochs, and willingly combine a lingering attachment to the ecological riches of the Holocene with responsiveness to the intense novelty of the Anthropocene, they will need a new conceptual vocabulary to steer by. Sustainability will start to look like a time-bound and contingent goal at best, not an absolute one, so environmentalists will need to construct some other normative standard of value. Otherwise, they run the risk of finding themselves fatalistically accepting environmental degradation as the inevitable outcome of the earth system's turbulence—when in fact many different paths forward are possible, and the chaotic nature of the crisis means that the flap of any given butterfly's wings might have a disproportionate influence on the new world that will eventually succeed the Holocene. We have seen that the idea of the Anthropocene can be a way of making connections between human and nonhuman power relations. Given that, the best conceptual repertoire for ecology at the birth of the Anthropocene might overlap a good deal with the vocabulary of democracy, devolution, and egalitarianism.

The rough transition out of the Holocene has seen the deadening or searing away of one ecological community after another. Its keynote environmental effects have been depatterning and subordination to single authorities. In the clear-cutting of forests; the draining of wetlands; the damming, dredging, channelization, and eutrophication of rivers; the exhaustion, salinization, contamination, and erosion of soil; bottom trawling; strip-mining; and the ever-expanding imposition of precarious, input-saturated monocultures, complex ecologies have been dispersed and simplified in order to tame them into servicing the extractive demands of international capital. To negotiate the transition in a just and bearable way, what is needed is to maximize the countervailing presence of plural, diverse, and polycentric ecosystems. Speaking very generally, the hardships of the transition can be cushioned by fostering variegated communities of life, dense with competing biogeochemical pathways. The most resilient ecosystems, just like the most flourishing civil societies, tend to be those that function as a conversation between many different voices.

A politics of the end-Holocene event could be devoted to nurturing evolved complexity and plurality and to resisting its subjection to the abstracting force of profit. Thus the idea of the Anthropocene is not by any means an alibi for anthropocentric resource extraction (as opponents of the concept have feared). Instead, the contemporary rewilding movement might offer a template for democratic ecological praxis at the birth of the Anthropocene, if that movement is driven by an attention to "process" rather than final outcomes, and if it deprecates attempts to manage the definitive forms that landscapes should take, preferring simply to reestablish and enhance "ecology's dynamic interactions" in all their variety.[3] More broadly, however, the struggle for an equitable path through the end-Holocene crisis will have to draw on the whole spectrum of initiatives already developed by environmental scholars and activists, from no-till farming to indigenous rights campaigns, and from distributed energy generation to participation in electoral politics. But the theory behind those initiatives must continue to evolve. The idea of the stratigraphic Anthropocene means demoting the old mantra of sustainability and putting in its place a more pragmatic first principle that is avowedly concerned with difference and plasticity instead of the defense of sameness. Against the flattening and simplifying impulse of the end-Holocene event, there is only this: the struggle for plural ecologies.

This suggests that the idea of the Anthropocene should tend more toward invigorating and even healing the environmental movement than toward exerting a corrective discipline over it. Still, it would be strange if such a fundamental reorientation of environmental principles, away from sustainability and toward a devolved ecological pluralism, did not lead to some significant practical recalibrations. The Anthropocene provides a way of seeing rather than a programmatic agenda, but any number of green strategies and principles might look different in light of the news from Mauna Loa and in the context of deep time. I suggest four possible changes of emphasis.

Firstly, thinking through the Anthropocene demands an avowedly global perspective. It means looking at the condition of the entire earth system and, hence, at the transnational effects of unequal relations of production and resource extraction. One important strand in green politics has been the anticonsumerist tradition, associated with nature conservation and quality of life issues, that emerged among affluent groups in the developed world—a tradition that must seem to many people the essence of what is meant by "environmentalism." But given its global perspective, an environmentalism for the birth of the Anthropocene might perceive its center of gravity as lying instead with the mass-based environmental justice movements of the majority part of the world.

The crisis accentuates rather than diminishes the differences between rich and poor. Among the dispossessed of the global South the social costs of export-led industrialization, toxic waste disposal, and the privatization of common land and fishing rights give rise to forms of collective resistance that often have an inescapably ecological dimension. Cartels of state officials and multinational corporations face challenges over flashpoints like pollution, illegal logging, and dam-building from broad antisystemic coalitions that draw on the contributions of small-scale farmers and fishers, women's movements, indigenous peoples, slum dwellers, and others whose livelihoods are assaulted by predatory development. This so-called environmentalism of the poor might appropriately become the type of environmentalism most readily associated with the idea of the Anthropocene.[4] That is not to downplay the classical contributions of western North America and its national parks, or West Germany's antinuclear culture and the electoral success of its Green Party. But the most important proving grounds for ecological ideas might increasingly be found elsewhere, in places where the costs of air and water pollution, soil degradation, and forest destruction are felt at their most raw. One might look to Latin America and to China, for instance, or to any of the regions—Southeast Asia from Kolkata to the Philippines, the Sahel, or the martyrized nations that make up the Alliance of Small Island States—that face the most extreme vulnerability to climate change.

The global is the local, as they say. The idea of the Anthropocene is necessarily cosmopolitan, but it urges us to examine the interplay between regional crises rather than to abstract them into any overarching planetary trajectory. In its systemic perspective, nearly all of humanity participates somehow in the end-Holocene event, but not as an undifferentiated mass: on the contrary, precisely through the differential ways in which societies and classes are incorporated into the ecological matrix of the global economy.

Secondly, the Anthropocene helps to make vivid the way in which human and nonhuman forces coproduce one another, so that any society is always composed from bundles of intrahuman and extrahuman relations. Opponents of the concept have often accused it of reimposing a dualistic split between humans and nature, and some scholars who have discussed the new epoch have given those opponents ammunition with their incautious references to the impact of "the human enterprise" on natural ecosystems. Conversely, one common theme among the proponents of the Anthropocene has been that the birth of the new epoch illustrates the absence of any metaphysical divide between humans and nature. True enough, although on its own that hardly explains what makes the concept so original. (Haven't those who write about the environment been ritually pointing out for decades that humans are also a part of nature?) The stratigraphic version of the Anthropocene offers a new twist on the old truism, however. Antonio Stoppani's and Jan Zalasiewicz's thought experiments about future alien geologists investigating the traces of human societies invite their readers to reimagine cities, roads, and cemeteries as incipient fossil assemblages. Doing so yields a bracing sense of how human labor participates ineluctably in much older ecological processes. For one example, take the observation in papers by members of the Anthropocene Working Group that the digging of mines is, at root, just a novel form of what geologists call "selective erosion."

The stratigraphic Anthropocene can put one on guard against philosophically clumsy dualism, and thereby it becomes a way of guarding against politically ugly universalism. The first step to making an undifferentiated mass of humankind the illusory protagonist of the Anthropocene is to give all humans something in common by setting them all in opposition to nature. The alternative is to define the characteristic force at work in the birth of the Anthropocene not as humankind but as *societies* that dialectically intertwine humans and others. Europe's Mesolithic foraging bands were assembled out of interactions between people, eels, flints, wild boars, hazelnuts, and hunting dogs. The Norte Chico civilization was made by anchovies, cotton, and reeds as well as by human hands. The idea of the Anthropocene helps us to see the societies of the end–Holocene event as still denser hybrids in which tangles of state power and class relations mediate the ecological relations between human and nonhuman forces.

Thirdly, the Anthropocene tends to underline the importance of modernity's biological economy, as opposed to the metal-and-concrete industrialization that is its most starkly visible calling card. Environmental degradation is nothing new, and nor are the social strains to which it gives rise. Intensive farming and timber-fueled manufacturing have helped to foster ecological crises for thousands of years. We saw in chapter 5 how the evolution of agriculture transformed large parts of the Holocene biosphere. Recalling this history draws attention to how food production has continued to shape the Anthropocene, while at the same time it underlines the fact that modern styles of farming, fishing, and forestry are just as novel and transformative as power stations and expressways. A rippling golden field of wheat is no less artificial than a factory floor. Holocene societies were shaped overwhelmingly by their diets, and the social life of food cannot be much less important to the Anthropocene, given that agriculture, forestry, and changes in land use presently contribute a quarter of all greenhouse gas emissions, and that farming takes up about a third of the total land area of the planet while more than eight hundred million people remain undernourished, by United Nations definitions.

Geologists of the future might first notice the Anthropocene as a biohorizon for species relocations, and on examining its strata they would be struck immediately by the bizarre makeup of its fossil pollen assemblages. Reflecting on the fossil traces that are currently being laid down directs attention to modernity's restructuring of the biosphere, and to the urgent need to restore some measure of ecological functioning to the agricultural deserts currently eroding away in a hail of fertilizer and pesticide. Thus the idea of the Anthropocene denormalizes industrial agriculture. It suggests that the routine practices of Smithfield Foods and Archer Daniels Midland should be as readily subject to environmentalist critique as Indonesian palm oil plantations or Monsanto's biotechnology patents. Agribusiness lies at the heart of the capitalist world-ecology. The U.S. National Grain and Feed Association, for instance, notes that its membership includes "commodity futures brokers; . . . banks; railroads; barge lines; grain exchanges; biotechnology providers; engineering and design/construction firms; insurance companies; computer/software firms; and other companies." And even fossil fuels themselves are organic products, of course. The single most important struggle at the birth of the Anthropocene is not to protect old-fashioned organic technologies against high-tech encroachment but precisely the reverse: to replace coal, oil, and gas with nonbiological sources of electricity.

That brings us to the last of these four examples of how the Anthropocene might change the emphasis of the environmental movement. Global warming is the central preoccupation of environmental politics, and the story of the Phanerozoic eon helps show why that should be the case. As we saw in chapter 4, changes of state in the earth system have extremely varied causes, but climate effects nearly always participate in them. From the mass extinction at the end of the Ordovician period to the Pleistocene glaciations, everything affects the climate system, and climate affects everything. At the center of this central system is the earth's paradigmatic biogeochemical loop: the carbon cycle. Carbon builds the whole biosphere, flows through the oceans, sediments, and atmosphere, and in its atmospheric phase decisively inflects the quantity and distribution of energy around the surface of the earth. Modern global warming has to be understood not as a deformation of the stable natural order but as the latest of the carbon cycle's ceaseless modulations.

By directing attention to fluctuations in the carbon cycle through geological time, the stratigraphic version of the Anthropocene encourages a change of focus from the demand side to the supply side of the fossil fuel industry. In geological terms, the real novelty is not a species maximizing the energy available to it but the recent scale of the "selective erosion" of fossil fuels, the mining and drilling processes that bring them to the surface. Recall the figures given in chapter 1: 240 billion metric tons of carbon accumulated in the atmosphere from 1750 to 2011, but a total indicated reserve of 780 billion tons of fossil-fuel carbon remains underground. The widespread concern with cutting back on the use of fossil fuels only tackles the issue indirectly. What ultimately matters is keeping most of that remaining carbon in the ground. The earth system concept of biogeochemical cycles illuminates the situation: carbon must pass through multiple phases in the course of its cycle. Traditionally, they can be grouped into biological and geological phases, but the carbon cycle now also includes a novelty: an *economic* phase. While fossil fuels are still underground, they become assets of the companies that own them. This economic phase is an essential part of the modern carbon cycle. If the reserves did not have economic value, labor power could not be mobilized to bring them to the surface; the quantity of value that they possess determines whether and when labor power is attracted to them.

The economic phase is not only the newest phase in carbon's biogeochemical cycle but also the one most vulnerable to disruption. If you want to interrupt the passage of carbon between subterranean and atmospheric reservoirs, the economic moment of the cycle, in which carbon is simultaneously underground and already present in balance sheets and stock market valuations, is the choke point that you might target. Only national governments have the strength to close that passage off, and they have strong incentives not to do so. But the carbon cycle's economic mechanism may be hampered by activist-led divestment from fossil fuel companies, which has the potential to accelerate cautious shareholders' retreat from firms whose value rests on their ownership of assets that may never be usable. The share prices of the largest fossil fuel reserve owners have become the critical variable of the planet's carbon cycle. Thus the idea of the Anthropocene offers a new context for understanding the fossil fuel divestment movement and, in particular, the way in which its organizers have sought to make visible the connections between the carbon dioxide readings at Mauna Loa and the institutions of global finance. Divestment from fossil fuels is a geological act, an intervention in the linchpin biogeochemical process of the earth system.

This book has sketched out the political implications of the version of the Anthropocene epoch that has been constructed since 2008 by the stratigraphers of the ICS Anthropocene Working Group. If a new epoch, in a formal geological sense—the thirty-ninth one of the Phanerozoic eon—may be said to have begun in 1952, then the world is effectively still in the midst of a transition between epochs: the end-Holocene event. Environmentalists' goal should not be to call off that transition and replace it with indefinite sustainability, but instead to intervene in it by guarding and rebuilding ecological pluralism. Construing the environmental crisis in these terms implies that the cutting edge of environmental praxis is the environmentalism of the poor in the global South, and that geological thought experiments can teach a lesson in how human societies really work; it implies that agribusiness is an indispensable target for environmentalist critique, and that fossil fuel divestment is a way of getting at the crux of how the planet's biogeochemical systems are changing.

The life of *Homo sapiens* now includes three epochs of geological time. The Pleistocene is the epoch of human biological evolution. It runs as far as the wide-eyed exploration of all six habitable continents, the masterpieces of Chauvet, Altamira, and Lascaux (standing in for countless others that are permanently lost or yet to be found), and the unspeakable tragedy of the megafaunal extinctions. The three ages of the Holocene make up the epoch in which class-stratified societies and domesticated species were coassembled over generations of field labor, the epoch in which farming, writing, the city, and both tyranny and democracy were born. The Anthropocene is the epoch still coming into being, as old assemblages are broken and new ones are formed at a dazzling rate. The Columbian exchange, the Industrial Revolution, the Great Acceleration: these and other inflection points involved a ramifying series of attachments between miscellaneous loci of dynamic activity. Networks formed; connections multiplied.

A history recounted in terms like these can foster both radical and conservative impulses. It describes a new world in the moment of its arrival, and in that way it asserts the pressing need to reimagine human and nonhuman life outside the confines of the Holocene. At the same moment, it poses the question of how best to keep faith with the web of relationships, dependencies, and symbioses that made up the planetary system of the dying epoch. The Holocene was not a backdrop to civilization but a mode of life in which the evolution of human societies participated. The coming mode of life will surely be no less rich than the old in ambivalences and imperfections, and the form that even its earliest stages will take is partly out of human control. But ongoing struggles for environmental justice still have the power to influence what will emerge from the Holocene's terminal crisis. If the planet is living through the birth of an epoch, then the world's green movements face the responsibility of helping shape a turning point within the vast reaches of the geological timescale.

I end with a kind of parable about what it means to bear witness to the birth of the Anthropocene, even though doing so goes against a principle of mine in this book and briefly lets one person's voice stand in for that of all humankind.

The nuclear bomb tests that provide the best marker for the coming of the new epoch involved more than just scientists and technicians. Investigating strategies for the nuclear battlefield, the U.S., Soviet, and British armies conducted training maneuvers amid the aftermath of the test shots. James Yeatts took part in the U.S. Army's Desert Rock exercises in 1952. Some years later he lost the teeth from his mouth, pulling them away with his bare hands. His son was born profoundly deformed. "When the bomb was detonated," he said, "we had our backs to the blast, kneeling with our hands over our eyes and our eyes closed. The flash was so bright we could see the bones in our hands."[5]

INTRODUCTION

1. Benjamin P. Horton et al., "Expert assessment of sea-level rise by AD 2100 and AD 2300," *Quaternary Science Reviews* 84 (2014): 1–6; Karen M. Robbirt et al., "Potential disruption of pollination in a sexually deceptive orchid by climatic change," *Current Biology* 24, no. 23 (2014): 2845–49; Philip Micklin, "The past, present, and future Aral Sea," *Lakes and Reservoirs: Research and Management* 15, no. 3 (2010): 193–213.

2. Don McKay, "Ediacaran and Anthropocene: Poetry as a reader of deep time," in *Making the Geologic Now*, ed. Elizabeth Ellsworth and Jamie Kruse (New York: Punctum, 2013), 46–54 (53).

CHAPTER ONE. LIVING IN DEEP TIME

1. "Heat-trapping gas passes milestone, raising fears," *New York Times*, May 10, 2013, www.nytimes.com. The story has become something of a touchstone in discussions of the Anthropocene: see Ian Baucom, "History 4°: Postcolonial method and Anthropocene time," *Cambridge Journal of Postcolonial Literary Inquiry* 1, no. 1 (2014): 123–42; Bruno Latour, "Agency at the time of the Anthropocene," *New Literary History* 45, no. 1 (2014): 1–18.

2. Chris Caseldine, "Conceptions of time in (paleo)climate science and some implications," *Wiley Interdisciplinary Reviews: Climate Change* 3, no. 4 (2012): 329–38 (334).

3. Stephen Jay Gould, *Time's Arrow, Time's Cycle: Myth and Metaphor in the Discovery of Geological Time* (Cambridge, MA: Harvard University Press, 1987), 3.

4. Matt Ridley, *The Rational Optimist* (London: HarperCollins, 2010), 329.

5. Colin Tudge, *The Day before Yesterday: Five Million Years of Human History* (London: Pimlico, 1996), 11–12, italics in the original.

6. John McPhee, *Encounters with the Archdruid* (New York: Farrar, Straus and Giroux, 1971), 79–80.

7. Roy Porter, *The Making of Geology: Earth Science in Britain, 1660–1815* (Cambridge: Cambridge University Press, 1977), 103, 142.

8. Martin J. S. Rudwick, *Worlds before Adam: The Reconstruction of Geohistory in the Age of Reform* (Chicago: University of Chicago Press, 2008), 558. What follows also draws on Rudwick's *Bursting the Limits of Time: The Reconstruction of Geohistory in the Age of Revolution* (Chicago: University of Chicago Press, 2005).

9. The neocatastrophist turn is distinct from (although it has prompted a fresh appreciation of) the original "catastrophism" of the seventeenth and eighteenth centuries. The catastrophist tradition was a loose genre of theories—all of which came to seem troublingly speculative within the gradualist orthodoxy—that explained the observed state of the planet by postulating one or more revolutionary upheavals, such as mountain-topping floods, in the distant past.

10. Hiroshi Miyake, Haruka Shibata, and Yasuo Furushima, "Deep-sea litter study using deep-sea observation tools," in *Interdisciplinary Studies on Environmental Chemistry*, vol. 5, ed. K. Omori et al. (Tokyo: Terrpub, 2011), 261–69; Maria Olech, "Human impact on terrestrial ecosystems in West Antarctica," *Proceedings of the NIPR Symposium on Polar Biology* 9 (1996): 299–306.

11. Vaclav Smil, "Harvesting the biosphere: The human impact," *Population and Development Review* 37, no. 4 (2011): 613–36.

12. William B. Whitman, David C. Coleman, and William J. Wiebe, "Prokaryotes: The unseen majority," *Proceedings of the National Academy of Sciences* 95, no. 12 (1998): 6578–83; Beth N. Orcutt et al., "Microbial ecology of the dark ocean above, at, and below the seafloor," *Microbiology and Molecular Biology Reviews* 75, no. 2 (2011): 361–422.

13. Les Watling and Elliott A. Norse, "Disturbance of the seabed by mobile fishing gear: A comparison to forest clearcutting," *Conservation Biology* 12, no. 6 (1998): 1180–97.

14. Jeremy B. C. Jackson, "Ecological extinction and evolution in the brave new ocean," *Proceedings of the National Academy of Sciences* 105, supplement 1 (2008): 11458–65 (11463); Daniel Pauly, Reg Watson, and Jackie Alder, "Global trends in world fisheries: Impacts on marine ecosystems and food security," *Philosophical Transactions of the Royal Society B* 360 (2005): 5–12.

15. Elizabeth Kolbert, *The Sixth Extinction: An Unnatural History* (London: Bloomsbury, 2014), 141.

16. J. R. McNeill, *Something New under the Sun: An Environmental History of the World in the 20th Century* (London: Penguin, 2000), 137, 243, 251–52; Daniel Pauly et al., "Towards sustainability in world fisheries," *Nature* 418 (2002): 689–95.

17. R. McLellan, L. Iyengar, B. Jeffries, and N. Oerlemans, eds., *Living Planet Report 2014* (Gland, Switzerland: World Wide Fund for Nature, 2014).

18. Erle C. Ellis, "Anthropogenic transformation of the terrestrial biosphere," *Philosophical Transactions of the Royal Society A* 369 (2011): 1010–35; Paul Robbins, *Lawn People* (Philadelphia, PA: Temple University Press, 2007), xiii.

19. Andrew Goudie, *The Human Impact on the Natural Environment: Past, Present and Future*, 7th ed. (Chichester, U.K.: Wiley-Blackwell, 2013), 172–73.

20. Bruce H. Wilkinson, "Humans as geologic agents: A deep-time perspective," *Geology* 33, no. 3 (2005): 161–64.

21. David Pimentel, "Soil erosion: A food and environmental threat," *Environment, Development and Sustainability* 8, no. 1 (2006): 119–37; United Nations Human Settlements Programme, *State of the World's Cities 2012/2013* (New York: Routledge, 2013), 151.

22. Andrew MacDowall, "War-torn and impoverished, Bosnia faces rebuild once again after floods," *Guardian*, May 26, 2014, www.theguardian.com.

23. Donald E. Canfield et al., "The evolution and future of Earth's nitrogen cycle," *Science* 330 (2010): 192–96; Yi Liu et al., "Global phosphorus flows and environmental impacts from a consumption perspective," *Journal of Industrial Ecology* 12, no. 2 (2008): 229–47.

24. Jonathan Safran Foer, *Eating Animals* (London: Penguin, 2010), 174.

25. Jeff Tietz, "Boss hog," *Rolling Stone*, December 14, 2006, www.rollingstone.com; Smithfield Foods, "*Rolling Stone*'s 'Bosshog' article: Fiction vs. fact," December 15, 2006, www.smithfieldfoods.com/newsroom/press-releases-and-news/rolling-stones-bosshog-article-fiction-vs-fact.

26. C. J. Moore et al., "A comparison of plastic and plankton in the North Pacific Central Gyre," *Marine Pollution Bulletin* 42, no. 12 (2001): 1297–1300; Thomas F. Stocker et al., eds., *Climate Change 2013: The Physical Science Basis*, IPCC Fifth Assessment Report (Cambridge: Cambridge University Press, 2013), 52. The latter is one of four volumes that make up the fifth report.

27. Anthony D. Barnosky et al., "Has the earth's sixth mass extinction already arrived?" *Nature* 471 (2011): 51–57. The extent of modern vertebrate extinction so far is assessed in Gerardo Ceballos et al., "Accelerated modern human-induced species losses: Entering the sixth mass extinction," *Science Advances* 1, no. 5 (2015), doi: 10.1126/sciadv.1400253.

28. *The IUCN Red List of Threatened Species*, www.iucnredlist.org/about /summary-statistics.

29. Kim Gehab et al., "Nile perch and the hungry of Lake Victoria: Gender, status and food in an East African fishery," *Food Policy* 33, no. 1 (2008): 85–98.

30. A. Rabatel et al., "Current state of glaciers in the tropical Andes: A multi-century perspective on glacier evolution and climate change," *The Cryosphere* 7, no. 1 (2013): 81–102; Fred Pearce, *With Speed and Violence* (Boston: Beacon, 2007), 182; Wolfgang Rack and Helmut Rott, "Pattern of retreat and disintegration of the Larsen B ice shelf, Antarctic Peninsula," *Annals of Glaciology* 39 (2004): 505–10.

31. Stocker et al., *Climate Change 2013*, 50–52.

32. Jeremy D. Shakun and Anders E. Carlson, "A global perspective on Last Glacial Maximum to Holocene climate change," *Quaternary Science Reviews* 29, no. 15–16 (2010): 1801–16; Shaun A. Marcott et al., "A reconstruction of regional and global temperature for the past 11,300 years," *Science* 339 (2013): 1198–1201.

33. Carbon Tracker Initiative, *Unburnable Carbon 2013: Wasted Capital and Stranded Assets* (2013), www.carbontracker.org; Christophe McGlade and Paul Ekins, "The geographical distribution of fossil fuels unused when limiting global warming to 2°C," *Nature* 517 (2015): 187–90. Seven hundred eighty billion metric tons of carbon is equivalent to 2860 billion tons of CO_2. That is the estimated quantity of the world's economically recoverable reserves; the total quantity of fossil fuel resources that could in principle be extracted at some point is far greater.

34. W. A. Kurz et al., "Mountain pine beetle and forest carbon feedback to climate change," *Nature* 452 (2008): 987–90.

35. Andrew H. MacDougall, Christopher A. Avis, and Andrew J. Weaver, "Significant contribution to climate warming from the permafrost carbon feedback," *Nature Geoscience* 5, no. 10 (2012): 719–21; NOAA National Climatic Data Center, www.ncdc.noaa.gov.

36. Mee Kam Ng, "World-city formation under an executive-led government: The politics of harbour reclamation in Hong Kong," *Town Planning Review* 77, no. 3 (2006): 311–37; Kenneth Pomeranz, "The great Himalayan watershed: Agrarian crisis, mega-dams and the environment," *New Left Review*, n.s., 58 (2009): 5–39.

CHAPTER TWO. VERSIONS OF THE ANTHROPOCENE

1. Christian Schwägerl, *The Anthropocene: The Human Era and How It Shapes Our Planet* (Santa Fe: Synergetic Press, 2014), 8–10; Fred Pearce, *With Speed and Violence* (Boston: Beacon, 2007), 21; Elizabeth Kolbert, *The Sixth Extinction: An Unnatural History* (London: Bloomsbury, 2014), 107–8.

2. Clive Hamilton and Jacques Grinevald survey these precursors, but stress the novelty of Crutzen's concept, in "Was the Anthropocene anticipated?" *Anthropocene Review* 2, no. 1 (2015): 59–72. On Buffon, see Noah Heringman, "Deep time at the dawn of the Anthropocene," *Representations* 129 (2015): 56–85.

3. Paul J. Crutzen and Eugene F. Stoermer, "The 'Anthropocene,'" *IGBP Newsletter*, no. 41 (2000): 16–18.

4. Paul J. Crutzen, "Geology of mankind," *Nature* 415 (2002): 23.

5. Will Steffen, Paul J. Crutzen, and John R. McNeill, "The Anthropocene: Are humans now overwhelming the great forces of nature?" *Ambio* 36, no. 8 (2007): 614–21. The graphs of the Great Acceleration are refined, extended, and thoughtfully reanalyzed in Will Steffen et al., "The trajectory of the Anthropocene: The Great Acceleration," *Anthropocene Review* 2, no. 1 (2015): 81–98.

6. William Ruddiman, "The anthropogenic greenhouse era began thousands of years ago," *Climatic Change* 61, no. 3 (2003): 261–93; Ruddiman, "The Anthropocene," *Annual Review of Earth and Planetary Sciences* 41 (2013): 45–68.

7. Andrew Glikson, "Fire and human evolution: The deep-time blueprints of the Anthropocene," *Anthropocene* 3 (2013): 89–92; Christopher E. Doughty, Adam Wolf, and Christopher B. Field, "Biophysical feedbacks between the Pleistocene

megafauna extinction and climate: The first human-induced global warming?" *Geophysical Research Letters* 37, no. 15 (2010): L15703; Bruce D. Smith and Melinda A. Zeder, "The onset of the Anthropocene," *Anthropocene* 4 (2013): 8–13; Giacomo Certini and Riccardo Scalenghe, "Anthropogenic soils are the golden spikes for the Anthropocene," *The Holocene* 21, no. 8 (2011): 1269–74.

8. Smith and Zeder, "Onset of the Anthropocene."

9. Dipesh Chakrabarty, "The climate of history: Four theses," *Critical Inquiry* 35, no. 2 (2009): 197–222 (212, 218).

10. Crutzen and Stoermer, "The Anthropocene," 18.

11. Chakrabarty, "Climate of history," 218, 221.

12. Will Steffen, Jacques Grinevald, Paul Crutzen, and John McNeill, "The Anthropocene: Conceptual and historical perspectives," *Philosophical Transactions of the Royal Society A* 369 (2011): 842–67 (843).

13. The most powerful indictment to date is Jason W. Moore, "The Capitalocene, part I: On the nature and origins of our ecological crisis" (2014), www .jasonwmoore.com. See also Slavoj Žižek, *Living in the End Times* (London: Verso, 2010), 327–36; Ben Dibley, "'Nature is us': The Anthropocene and species-being," *Transformations* 21 (2012); and Christophe Bonneuil's discerning typology of viewpoints on the new epoch, "The geological turn: Narratives of the Anthropocene," in *The Anthropocene and the Global Environmental Crisis: Rethinking Modernity in a New Epoch*, ed. Clive Hamilton, Christophe Bonneuil, and François Gemenne (Abingdon, U.K.: Routledge, 2015), 17–31. Vigorous presentations of the main objections include Eileen Crist, "On the poverty of our nomenclature," *Environmental Humanities* 3 (2013): 129–47; George Wuerthner, Eileen Crist, and Tom Butler, eds., *Keeping the Wild: Against the Domestication of Earth* (Washington, DC: Island Press, 2014); John Lewin and Mark G. Macklin, "Marking time in geomorphology: Should we try to formalise an Anthropocene definition?" *Earth Surface Processes and Landforms* 39, no. 1 (2014): 133–37; Andreas Malm and Alf Hornborg, "The geology of mankind? A critique of the Anthropocene narrative," *Anthropocene Review* 1, no. 1 (2014): 62–69; and Jeremy Baskin, "Paradigm dressed as epoch: The ideology of the Anthropocene," *Environmental Values* 24, no. 1 (2015): 9–29.

14. Gerda Roelvink, "Rethinking species-being in the Anthropocene," *Rethinking Marxism* 25, no. 1 (2013): 52–69 (53).

15. Chakrabarty, "Climate of history," 206.

16. Chakrabarty, "Postcolonial studies and the challenge of climate change," *New Literary History* 43, no. 1 (2012): 1–18 (11, 14).

17. See also Nigel Clark, "Rock, life, fire: Speculative geophysics and the Anthropocene," *Oxford Literary Review* 42, no. 2 (2012): 259–76.

18. Dipesh Chakrabarty, "Climate and capital: On conjoined histories," *Critical Inquiry* 41, no. 1 (2014): 1–23.

19. Jan Zalasiewicz, *Anthropocene Working Group, Newsletter 1* (December 2009), quaternary.stratigraphy.org/workinggroups/anthropocene.

20. Strictly speaking, the International Chronostratigraphic Chart is a representation of two parallel time schemes in one. Chronostratigraphy is concerned with the relative dating of stratified rock layers, which is a concept that differs somewhat from the sequencing of geological time per se. Chronostratigraphic "time-rock" units and geological time units correspond to one another exactly, however. The subtle distinction between them will not trouble us here.

21. International Commission on Stratigraphy, "GSSP table for Rupelian stage," www.stratigraphy.org/GSSP/Rupelian.html.

22. Jan Zalasiewicz and Mark Williams, "Letter to potential members," quaternary.stratigraphy.org/workinggroups/anthropocene.

23. Jan Zalasiewicz et al., "Are we now living in the Anthropocene?" *GSA Today* 18, no. 2 (2008): 4–8.

CHAPTER THREE. GEOLOGY OF THE FUTURE

1. Bronislaw Szerszynski, "The end of the end of nature: The Anthropocene and the fate of the human," *Oxford Literary Review* 34, no. 2 (2012): 165–84.

2. Jan Zalasiewicz et al., "Response to Autin and Holbrook on 'Is the Anthropocene an issue of stratigraphy or pop culture?'" *GSA Today* 22 (2012), online only, www.geosociety.org/gsatoday. See also Ian Angus, "When did the Anthropocene begin . . . and why does it matter?" *Monthly Review* 67, no. 4 (2015), monthlyreview.org.

3. Colin N. Waters et al., "A stratigraphical basis for the Anthropocene?" in *A Stratigraphical Basis for the Anthropocene*, ed. C. N. Waters et al. (London: Geological Society, 2014).

4. Etienne Turpin and Valeria Federighi, "A new element, a new force, a new input: Antonio Stoppani's Anthropozoic," in *Making the Geologic Now*, ed. Elizabeth Ellsworth and Jamie Kruse (New York: Punctum, 2013), 40.

5. The principal sources for what follows are Jan Zalasiewicz, *The Earth after Us: What Legacy Will Humans Leave in the Rocks?* (Oxford: Oxford University Press, 2008), and Waters et al., *A Stratigraphical Basis for the Anthropocene*.

6. Peter M. Vitousek et al., "Biological invasions as global environmental change," *American Scientist* 84, no. 5 (1996): 468–78.

7. Laura D. Triplett et al., "The potential for multiple signatures of invasive species in the geologic record," *Anthropocene* 5 (2014): 59–64; H. A. Mooney and E. E. Cleland, "The evolutionary impact of invasive species," *Proceedings of the National Academy of Sciences* 98, no. 10 (2001): 5446–51; David L. Strayer et al., "Understanding the long-term effects of species invasions," *Trends in Ecology and Evolution* 21, no. 11 (2006): 645–51.

8. David Archer, *The Long Thaw* (Princeton, NJ: Princeton University Press, 2009), 156.

9. Thomas Sumner, "No stopping the collapse of West Antarctic Ice Sheet," *Science* 344 (2014): 683.

10. Jan Zalasiewicz, Colin N. Waters, and Mark Williams, "Human bioturbation, and the subterranean landscape of the Anthropocene," *Anthropocene* 6 (2014): 3–9.

11. Zalasiewicz, *Earth after Us*, 189.

12. Ibid., 236–38.

13. S. C. Finney, "The 'Anthropocene' as a ratified unit in the ICS International Chronostratigraphic Chart: Fundamental issues that must be addressed by the Task Group," in C. N. Waters et al., *A Stratigraphical Basis*, 26.

14. Kevin J. R. Rosman et al., "Lead from Carthaginian and Roman Spanish mines isotopically identified in Greenland ice dated from 600 B.C. to 300 A.D." *Environmental Science and Technology* 31, no. 12 (1997): 3413–16.

15. See, however, Kent G. Lightfoot et al., "European colonialism and the Anthropocene: A view from the Pacific coast of North America," *Anthropocene* 4 (2013): 101–15; Simon L. Lewis and Mark A. Maslin, "Defining the Anthropocene," *Nature* 519 (2015): 171–80.

16. Anthony D. Barnosky et al., "Prelude to the Anthropocene: Two new North American Land Mammal Ages (NALMAs)," *Anthropocene Review* 1, no. 3 (2014): 225–42.

17. Jason W. Moore, "Ecology and the rise of capitalism" (PhD diss., University of California, Berkeley, 2007), www.jasonwmoore.com.

18. M. Williams et al., "Is the fossil record of complex animal behaviour a stratigraphical analogue for the Anthropocene?" in C. N. Waters et al., *A Stratigraphical Basis*, 143–48.

19. Victoria C. Smith, "Volcanic markers for dating the onset of the Anthropocene," in C. N. Waters et al., *A Stratigraphical Basis*, 283–99.

20. International Monetary Fund, *World Economic Outlook: Asset Prices and the Business Cycle, May 2000* (Washington: IMF, 2000), 154; B. A. Holderness, *British Agriculture since 1945* (Manchester, U.K.: Manchester University Press, 1985), 113–15.

21. J. R. McNeill, *Something New under the Sun: An Environmental History of the World in the 20th Century* (London: Penguin, 2000).

22. Anthony D. Barnosky, "Palaeontological evidence for defining the Anthropocene," in C. N. Waters et al., *A Stratigraphical Basis*, 155, 161.

23. I. P. Wilkinson et al., "Microbiotic signatures of the Anthropocene in marginal marine and freshwater palaeoenvironments," in C. N. Waters et al., *A Stratigraphical Basis*, 206.

24. Alexander P. Wolfe et al., "Stratigraphic expressions of the Holocene-Anthropocene transition revealed in sediments from remote lakes," *Earth-Science Reviews* 116 (2013): 17–34; Donald E. Canfield et al., "The evolution and future of Earth's nitrogen cycle" *Science* 330 (2010): 192–96.

25. Graeme T. Swindles et al., "Spheroidal carbonaceous particles are a defining stratigraphic marker for the Anthropocene," *Scientific Reports* 5, no. 10264 (2015), doi: 10.1038/srep10264; Neil L. Rose, "Spheroidal carbonaceous fly ash particles provide a globally synchronous stratigraphic marker for the Anthropocene," *Environmental Science and Technology* 49, no. 7 (2015): 4155–62.

26. Jan Zalasiewicz et al., "When did the Anthropocene begin? A mid-twentieth-century boundary level is stratigraphically optimal," *Quaternary*

International 383 (2015): 196–203 (200). Mountain War Time was six hours behind Greenwich Mean Time.

27. Mike Walker, Phil Gibbard, and John Lowe, "Comment on 'When did the Anthropocene begin? A mid-twentieth-century boundary is stratigraphically optimal,' by Jan Zalasiewicz et al." *Quaternary International* 383 (2015): 204–7.

28. Gary J. Hancock et al., "The release and persistence of radioactive anthropogenic nuclides," in C. N. Waters et al., *A Stratigraphical Basis*, 265–81; Colin N. Waters et al., "Can nuclear weapons fallout mark the beginning of the Anthropocene Epoch?" *Bulletin of the Atomic Scientists* 71, no. 3 (2015): 46–57. The first thermonuclear explosion was the Ivy Mike test shot, which took place on—and utterly destroyed—the Pacific coral island of Elugelab. Its detonation at 07:14:59.4 (± 0.2 s) local time on November 1, 1952, perhaps provides an alternative option for an Anthropocene GSSA. Richard Rhodes, *Dark Sun: The Making of the Hydrogen Bomb* (New York: Simon and Schuster, 1995) is a detailed account.

29. Lewis and Maslin, "Defining the Anthropocene"; Jan Zalasiewicz et al., "Colonization of the Americas, 'Little Ice Age' climate, and bomb-produced carbon: Their role in defining the Anthropocene," *Anthropocene Review* 2, no. 2 (2015): 117–27.

CHAPTER FOUR. THE RUNGS ON THE LADDER

1. The narrative that follows draws on numerous sources. Of the surveys of earth history that I consulted, I am most indebted to Steven M. Stanley, *Earth System History*, 3rd ed. (New York: W. H. Freeman, 2009). Figures for extinction rates during the "Big Five" mass extinctions are taken from Anthony D. Barnosky et al., "Has the earth's sixth mass extinction already arrived?" *Nature* 471 (2011): 51–57.

2. Paul F. Hoffman and Daniel P. Schrag, "Snowball Earth," *Scientific American* (January 2000): 68–75.

3. Stephen Jay Gould, *Wonderful Life: The Burgess Shale and the Nature of History* (London: Vintage, 2000), 38.

4. Richard G. Klein, "Archaeology and the evolution of human behaviour," *Evolutionary Anthropology* 9, no. 1 (2000): 17–36.

5. John J. Shea, "*Homo sapiens* is as *Homo sapiens* was: Behavioral variability versus 'behavioral modernity' in Paleolithic archaeology," *Current Anthropology* 52, no. 1 (2011): 1–35.

6. Sally McBrearty and Alison S. Brooks, "The revolution that wasn't: A new interpretation of the origin of modern human behaviour," *Journal of Human Evolution* 39, no. 5 (2000): 453–563 (528–31).

7. Marlize Lombard and Isabelle Parsons, "What happened to the human mind after the Howiesons Poort?" *Antiquity* 85 (2011): 1433 43.

8. Sonia Harmand et al., "3.3-million-year-old stone tools from Lomekwi 3, West Turkana, Kenya," *Nature* 521 (2015): 310–15.

9. Daniel Lord Smail, *On Deep History and the Brain* (Berkeley: University of California Press, 2008), 190–91.

10. A shrinking minority of scholars do advocate stripping the Holocene of its status as an epoch and instead calling it the Flandrian age (an age is the unit hierarchically below an epoch).

11. Mark Williams et al., "The Anthropocene biosphere," *Anthropocene Review* 2, no. 3 (2015): 196–219.

CHAPTER FIVE. AN OBITUARY FOR THE HOLOCENE

1. James E. Hansen and Makiko Sato, "Paleoclimate implications for human-made climate change," in *Climate Change: Inferences from Paleoclimate and Regional Aspects*, ed. André Berger, Fedor Mesinger, and Djordje Šijački (Vienna: Springer, 2012), 21–47 (37, 39).

2. Shaun A. Marcott et al., "A reconstruction of regional and global temperature for the past 11,300 years," *Science* 339 (2013): 1198–1201; H. Renssen, H. Goosse, T. Fichefet, and J.-M. Campin, "The 8.2 kyr BP event simulated by a global atmosphere–sea–ice–ocean model," *Geophysical Research Letters* 28, no. 8 (2001): 1567–70.

3. The most significant of all these papers was Gerard Bond et al., "A pervasive millennial-scale cycle in North Atlantic Holocene and Glacial climates," *Science* 278 (1997): 1257–66.

4. The most important general sources for the account that follows are David E. Anderson, Andrew S. Goudie, and Adrian G. Parker, *Global Environments through the Quaternary: Exploring Environmental Change* (Oxford: Oxford University Press, 2007); William Burroughs, *Climate Change in Prehistory: The End of the Reign of Chaos* (Cambridge: Cambridge University Press, 2005); Jared Diamond, *Guns, Germs and Steel: A Short History of Everybody for the Last 13,000 Years* (London:

Vintage, 1998); Brian Fagan, *The Long Summer: How Climate Changed Civilization* (New York: Basic Books, 2004); Steven J. Mithen, *After the Ice: A Global Human History, 20,000-5000 BC* (London: Phoenix, 2004); Chris Scarre, ed., *The Human Past: World Prehistory and the Development of Human Societies*, 3rd ed. (London: Thames and Hudson, 2013); and the early volumes of *The Cambridge World History*, gen. ed. Merry E. Wiesner-Hanks (Cambridge: Cambridge University Press, 2015). For current thinking on agriculture's emergence in the Fertile Crescent and in China, I have generally followed Melinda A. Zeder, "The origins of agriculture in the Near East" and David Joel Cohen, "The beginnings of agriculture in China: A multiregional view," both in *Current Anthropology* 52, supplement 4 (2011): S221-35, S273-93.

5. Paul L. Koch and Anthony D. Barnosky, "Late Quaternary extinctions: State of the debate," *Annual Review of Ecology, Evolution, and Systematics* 37 (2006): 215-50.

6. N. Ray and J. M. Adams, "A GIS-based vegetation map of the world at the Last Glacial Maximum (25,000-15,000 BP)," *Internet Archaeology* 11 (2001), intarch.ac.uk.

7. Jerome E. Dobson, "Aquaterra *incognita:* Lost land beneath the sea," *Geographical Review* 104, no. 2 (2014): 123-38.

8. Julian B. Murton et al., "Identification of Younger Dryas outburst flood path from Lake Agassiz to the Arctic Ocean," *Nature* 464 (2010): 740-43. The flood had previously been thought to have traveled east down the Saint Lawrence into the Atlantic.

9. Mike Walker et al., "Formal definition and dating of the GSSP (Global Stratotype Section and Point) for the base of the Holocene using the Greenland NGRIP ice core, and selected auxiliary records," *Journal of Quaternary Science* 24, no. 1 (2009): 3-17.

10. Some hunter-gatherer groups did face the same hardships, however, because processing certain staple foraged foods—notably acorns—can be equally backbreaking work. And many East Asian societies developed pottery before agriculture, enabling grain cultures that were based on boiling and steaming rather than grinding and baking.

11. Lesley Head, "Contingencies of the Anthropocene: Lessons from the Neolithic" (*Anthropocene Review* 1, no. 2 [2014]: 113-25), is an astute account of

another way in which it is worthwhile to juxtapose the Anthropocene with the agricultural or Neolithic revolution.

12. Brian F. Byrd, "Public and private, domestic and corporate: The emergence of the Southwest Asian village," *American Antiquity* 59, no. 4 (1994): 639–66.

13. As proposed in Ian Hodder and Craig Cessford, "Daily practice and social memory at Çatalhöyük," *American Antiquity* 69, no. 1 (2004): 17–40.

14. Patrick Lajeunesse and Guillame St-Onge, "The subglacial origin of the Lake Agassiz–Ojibway final outburst flood," *Nature Geoscience* 1, no. 3 (2008): 184–88; David W. Leverington, Jason D. Mann, and James T. Teller, "Changes in the bathymetry and volume of Glacial Lake Agassiz between 9200 and 7700 ^{14}C yr B.P.," *Quaternary Research* 57, no. 2 (2002): 244–52.

15. Bryony J. Coles, "Doggerland: The cultural dynamics of a shifting coastline," in *Coastal and Estuarine Environments: Sedimentology, Geomorphology and Geoarchaeology*, ed. K. Pye and J. R. L. Allen (London: Geological Society, 2000), 393–401 (399).

16. Nick Brooks, "Cultural responses to aridity in the Middle Holocene and increased social complexity," *Quaternary International* 151, no. 1 (2006): 29–49.

17. Peter deMenocal et al., "Abrupt onset and termination of the African Humid Period: Rapid climate responses to gradual insolation forcing," *Quaternary Science Reviews* 19, no. 1–5 (2000): 347–61 (355).

18. Sing C. Chew, *World Ecological Degradation: Accumulation, Urbanization, and Deforestation, 3000 B.C.–A.D. 2000* (Walnut Creek, CA: AltaMira, 2001), 61.

19. Eric H. Cline, *1177 B.C.: The Year Civilization Collapsed* (Princeton, NJ: Princeton University Press, 2014).

20. Edmund Burke III, "The transformation of the Middle Eastern environment, 1500 B.C.E.–2000 C.E." (in *The Environment and World History*, ed. Edmund Burke III and Kenneth Pomeranz [Berkeley: University of California Press, 2009], 81–117) assesses the acute relative decline that the heartland of Holocene Afro-Eurasia underwent in the course of the transition to the Anthropocene.

21. M. J. C. Walker et al., "Formal subdivision of the Holocene Series/Epoch: A discussion paper by a Working Group of INTIMATE (Integration of ice-core, marine and terrestrial records) and the Subcommission on Quaternary Stratigraphy (International Commission on Stratigraphy)," *Journal of Quaternary Science* 27, no. 7 (2012): 649–59.

CONCLUSION

1. The most notable of several exceptions to this rule is the work of Peter Haff. See his "Technology as a geological phenomenon: Implications for human well-being," in *A Stratigraphical Basis for the Anthropocene*, ed. C. N. Waters et al. (London: Geological Society, 2014), 301–9.

2. Andrew Dobson, *Green Political Thought*, 4th ed. (Abingdon, U.K.: Routledge, 2007); Timothy Mitchell, *Carbon Democracy: Political Power in the Age of Oil* (London: Verso, 2013).

3. George Monbiot, *Feral: Rewilding the Land, Sea and Human Life* (London: Penguin, 2014), 83–84.

4. Gaia Vince's *Adventures in the Anthropocene: A Journey to the Heart of the Planet We Made* (London: Chatto and Windus, 2014) is the outstanding work so far to have linked the theme of the Anthropocene to detailed reportage on the environmentalism of the poor and its uncertain prospects.

5. Gerard DeGroot, *The Bomb: A History of Hell on Earth* (London: Pimlico, 2005), 242.

anomalocaridids, 116, 117
Antarctica, 30, 38, 40, 79–80, 107, 124, 128
Anthropocene: five maxims about, 108–10; "good," 197; implications for environmental politics, 4–6, 13–14, 84, 193–209; introduction of term, 42–44; popularization of term, 6, 51–52; as shock tactic, 70, 194–95; significance of name, 6, 10, 42, 52, 70–76, 195; varied meanings of, 6, 41, 44–48, 52, 55–56, 63–66
Anthropocene epoch: and anthropocentrism, 7, 74–76; and dualism, 7–8, 49–50, 54, 73–74, 76, 83, 109, 151, 203–4; and ecological pluralism, 6, 194, 201–2, 208; and remote future, 66–67, 76–85, 88, 105–6, 142–43, 204; and technocracy, 8, 52–55, 104 (*see also* geoengineering); and universalism, 7, 50–59, 61–62, 71, 76, 109, 160–61, 195, 204
Anthropocene epoch, dating of, 43, 45–48, 67–68, 74–76, 84–87, 89–108; use of 1952 as starting date, 105–8, 190, 197, 208
Anthropocene Review, 52
Anthropocene Working Group, 69, 71, 84–85, 208; foundation of, 64–65; and subdivision of Holocene, 190–91; writings by members of, 74, 75, 104, 195, 204. *See also* Zalasiewicz, Jan
antimony, 103
antiquarian history, 27
Appalachian Mountains, 117
aquaterra, 153–56
Aral Sea, 1
Archaeopteris, 118–19, 134
Archer Daniels Midland, 205
Arctic, 105, 131, 157, 158, 169; climate of, 19, 110, 127

Asia: Central, 153, 174, 177, 181, 184; East, 93–94, 178; Northeast, 164; South, 176, 181; Southeast, 176, 177, 184, 203; Southwest, 135, 157–59, 162–86. *See also individual countries*
Asia, pre-Holocene, 127, 128, 129, 152, 153, 156
Assyria, 178–80, 181
Atlantic islands, 93, 95
Atlantic Meridional Overturning Circulation, 36, 130–32, 157, 158–59, 166, 169, 189
Atlantic Ocean, 17, 129, 130, 147; North, 32, 67, 184; South, 40
Aurignacian culture, 85
Australia: contemporary environment, 50, 78, 79; history of, 124, 128, 135, 152, 153, 156, 170, 178
Avalonia, 118, 119
Aztec empire, 186, 187

Babylonia, 170, 177
Balkans, 34
balloon vine, 79
Baltica, 118
Bantu-speaking peoples, 178, 182, 183–84
barley, 157, 163, 165, 171–72, 175, 177, 179, 182
Barnosky, Anthony, 93, 102–3
Battle of Hastings, 21
Battle of the Boyne, 22
Beatles, 106, 107
behavioral modernity, 135–38
Beidha, 165–66
Beringia, 127, 156, 173, 187
Bhopal, 63
Bible, 25, 40
Big Bang, 22
biological pump, 60, 132

biomass, global, 31–32
Black Death, 96, 185–86
Black Sea flood, 169
Bolivia, 38
Botai culture, 175
Brazil, 18, 40, 95
Bretton Woods system, 199
Britain, 22, 78, 100; in Holocene, 170, 172, 179; and Industrial Revolution, 95–97, 98; and Last Glacial Maximum, 153; nuclear program, 105, 209
Bronze Age collapse, 179–80
bronze working, 171, 177
Brower, David, 25–26, 84, 139, 149, 195
Buffon, Georges-Louis Leclerc, comte de, 43
burial, 82, 135, 137, 168, 171
Byzantine empire, 183, 184, 186

C_4 grasses, 129, 133–34
Cage, John, 107
Cambrian period, 116–17, 125
Canada, 38, 78, 153, 158, 169
Canary Islands, 93, 95
capitalism: Dipesh Chakrabarty on, 48–51, 57; in Columbian exchange, 94–95, 186–87; in Industrial Revolution, 96–97; and significance of Anthropocene, 62, 107, 192, 196–97; and sustainability, 198
Capitalocene, 52, 94
carbon bubble, 38–39, 207, 214n33
carbon dioxide: in Holocene, 17, 168; and Little Ice Age, 94; in Phanerozoic eon, 18, 117–32; recent rise in concentrations, 16–19, 38–40, 147, 196, 206–8; and snowball earth, 114; used to define Anthropocene epoch, 43, 46–47, 98–99, 102

Carboniferous period, 71–72, 119, 122, 125
carbon weathering, 59–60, 117–18, 119, 129
Caribbean, 32, 95, 96
Caribbean monk seal, 32
Carthage, 91
Caseldine, Chris, 21
cassava, 93
Castile, 186–87
Çatalhöyük, 167–68, 188
catastrophism, 212n9
cattle, 31, 92, 165, 166, 167, 175, 177, 182
Cenozoic era, 125–33, 140, 144, 146
Chad, 139, 161
Chakrabarty, Dipesh, 48–51, 57–61, 63, 107, 118
Chauvet cave, 136, 140, 208
Chew, Sing, 179
Chicxulub bolide, 21, 29, 102, 124–25
Chile, 156, 158
China: pre-imperial, 158, 164, 168, 172–73, 176, 177, 180; imperial, 92–93, 94, 96–97, 98, 162, 181–82, 184, 185–86; contemporary, 40, 203
cities, 34, 40, 81–82, 102, 174–75, 179
climate, relative stability of in Holocene, 146–47, 149, 159, 195–96
climate change, anthropogenic, 16–19, 38–40, 43, 79, 147, 206–8; Dipesh Chakrabarty on, 48–51, 57; conspiracy theories about, 23–24, 26; future prospects, 79–80, 142, 147, 197; preindustrial, 46, 94
climate change, nonanthropogenic: in Holocene, 149–51, 164, 166, 168–70, 173–75, 176–77, 178, 184, 189–90; in late Pleistocene, 38, 153–58; in Phanerozoic eon, 117–32; rapidity of, 9–10, 29, 127, 146, 159; snowball earth, 114–15

coal, 38–39, 71–72, 96, 97–98, 119, 206
cod, 67, 92
Colombia, 38
Colorado potato beetle, 73
Columbian exchange, 91–95, 196, 208–9
Columbus, Christopher, 92, 107
common reed, Eurasian, 78
Congress of Vienna, 99
Constantinople, 163, 183, 186
Copenhagen, University of, 159
Copernicus, Nicolaus, 70–71
coral, 32, 33, 79, 102, 117, 119, 124,
    220n28; uses in stratigraphy, 79, 81,
    102, 106, 192
Cordilleran ice sheet, 156
corn, 80, 93, 103, 180, 185
Cosmoscene, 52, 70
cotton, 92, 96–97, 178
Cretaceous period, 72, 77, 102, 124–25
Crutzen, Paul, 12, 42–45, 46, 48, 55, 70,
    94, 95, 99; and Dipesh Chakrabarty,
    49–51; coins term *Anthropocene*, 42,
    142; and geoengineering, 54; and
    stratigraphic Anthropocene, 63, 65,
    67, 71, 77, 90
Cryogenian period, 72, 191
Cuban missile crisis, 105
Cuvier, Georges, 28
Cyprus, 165

Darwin, Charles, 28, 137
deforestation: and global biomass, 31;
    stratigraphic implications of, 80, 91,
    92, 98
Denisova Cave, 138
Desert Rock exercises, 209
Devonian period, 74, 118–19, 125
dingo, 178
dinosaurs, 123, 124–25, 126–27; extinc-
    tion of nonavian, 21, 29, 124–25

Diprotodon, 152
disease, 92, 96, 160, 187
divestment, from fossil fuels, 207–8
D'mt, 182
Dobson, Andrew, 199
dog, domesticated, 165
Doggerland, 164, 170, 190
domestication, 47, 136, 157, 158; in
    Holocene, 163–68, 171–73, 177, 208
Dubai, 85

early spider orchid, 1
East India Company, 96–97
Ediacaran biota, 11, 115, 145, 193
Eldredge, Niles, 29
Econocene, 52, 70
Economist (London), 51
ecosystem-services markets, 54
Ecuador, 164
Egypt: Dynastic, 175, 176–77, 179–80,
    189–90; 1952 revolution, 107, 190,
    197; Ottoman conquest, 186
Ellison, Ralph, 107
Elugelab, 220n28
end-Holocene event: characterisation of,
    90, 107, 162–63, 186–87, 196–97,
    200–205; as geological concept,
    94–95, 192, 196, 208
"environmentalism of the poor," 202–3,
    208
Eocene epoch, 19, 127–28
epoch, geological, ambivalent meaning
    of, 140–44
Eritrea, 139
erosion: of agricultural soil, 34, 91, 93,
    170, 179, 201, 205; as geological
    process, 29, 59–60, 123; of strati-
    graphic sites, 67, 77, 82, 88
Euphrates, 162, 165, 171, 181
Eurocentrism, 95–96, 101, 104, 135

Europe: in Columbian exchange and Industrial Revolution, 92–99; extinctions in, 152; geological science in, 27–28; geology and geography, 128, 129, 162; in Holocene, 164, 169, 170, 172, 175, 177, 184, 185–86; and human evolution, 135, 137; and Last Glacial Maximum, 38, 153; and Late Glacial, 156–57, 158. *See also* imperialism, European; *and individual countries and regions*

European Union Emissions Trading System, 54

extinction: in Great American Interchange, 130; of Pleistocene megafauna, 19, 47, 132, 141, 152–53, 156, 208; recent and prospective, 19, 36–37, 78, 89, 110, 142

extinctions, "Big Five," 36, 133; end-Cretaceous, 124–25 (*see also* Chicxulub bolide); end-Ordovician, 117–18, 206; end-Permian, 95, 122–23, 127; end-Triassic, 123–24; Late Devonian, 118–19

fall from Eden, 7, 25, 108, 161

Faulkner, William, 194

feedback mechanisms, 9, 38, 39

Fertile Crescent, 162, 223n20. *See also* Asia: Southwest

Finney, Stanley, 88

fishing, modern industrial, 32–33, 102, 205

Flandrian age, 221n10

Folsom culture, 166

foraging societies. *See* hunter-gatherer societies

France, 18; French Revolution, 17, 27, 28

future, remote, 66–67, 76–85, 88, 105–6, 142–43, 204

gazelle, 157, 163, 167

geoengineering, 44, 54, 89, 197–98

Geological Society of London, 65, 91

geological timescale: diagram, 4; principles of, 3, 64–65, 85–88, 112; status of Holocene epoch in, 140–44, 188–92, 221n10; structure of, 115, 122, 125–26; "time-rock" units, 217n20; unfamiliarity to nonspecialists, 20–23

Germany, 98, 153, 203

Gilgal, 163, 192

glaciers. *See* ice

Global Boundary Stratotype Section and Point (GSSP): for Anthropocene, 91, 93, 97–99, 102–4, 105–7, 197; for Holocene epoch and ages, 159, 189–91; requirements for, 86–87

globalization, 48–49, 57, 62, 91–97, 100–101. *See also* imperialism, European

Global Standard Stratigraphic Age (GSSA), 86, 91, 93, 95, 104–5, 220n28

global warming, as term 56, 57. *See also* climate change, anthropogenic; climate change, nonanthropogenic

Globo, O (Rio de Janeiro), 18

goats, 165, 167, 171–72, 175–76, 178

Göbekli Tepe, 164, 165, 168

golden spike. *See* Global Boundary Stratotype Section and Point (GSSP)

Gondwanaland, 117–19, 124

Gould, Stephen Jay, 23, 29, 115–16

gradualism, 29

Great Acceleration, 45, 100–101, 197, 208–9

"Great Dying," 122–23

Great Oxygenation Event, 60

great Pacific garbage patch, 36

Greece, 179, 181
Greenland: and modern climate change, 36, 80; past climate of, 153, 157, 158, 169; study of ice cores from, 91, 103, 157, 159, 189
Guardian (London), 18

Hadley cells, 63
Haff, Peter, 224n1
Hansen, James, 146, 149, 196
Hawaii, 78, 79, 184. *See also* Mauna Loa Observatory
Hercynian Mountains, 119
Himalayan Mountains, 129, 176
Hiroshima, 104
Holocene Climatic Optimum, 38, 166, 168, 170
Holocene epoch: agriculture and, 159-61; Paul Crutzen on, 42; division into geological ages, 188-91; GSSP (golden spike) for, 159; history of, 163-87; partisanship for, 146-49, 192, 200; relation to Anthropocene epoch, 5-6, 47, 88-90, 145-51, 161, 187-88, 191-92, 209; William Ruddiman on, 46; status as epoch, 140-44, 221n10
Homer, 181
Homo ergaster, 139, 140
Homo floresiensis, 138
Homogenocene, 52
Homo neanderthalensis, 135, 137, 139, 152
Hong Kong, 40
horses, 92, 93, 100, 175
human evolution, 134-40
Humboldt Current, 178
hunter-gatherer societies: and evolution of agriculture, 159-61; in Holocene, 92, 163-70, 172, 173, 175, 177, 178, 182-3, 222n10; in Pleistocene, 153,

157-58; as pre-Holocene mode of subsistence, 22, 146-47, 195-96

ice: in late Pleistocene and Holocene, 153, 157, 158, 164, 170; and modern climate change, 36, 38, 75-76, 79-80, 107, 142; in Phanerozoic eon, 19, 21, 117-18, 119, 122, 128, 131-32; as stratigraphic record, 77, 98, 103, 106, 159, 189; snowball earth, 114-15
imperialism, European, 17, 48-49, 53, 91-97, 186-87
Inca empire, 186, 187
India: geology, 124, 128, 129, 156; in Holocene, 176, 177, 182, 190; modern, 94, 96-97, 162
Indian Ocean, 93-94, 95, 162, 183-84, 186
Indohyus, 128
Indonesia, 156, 180; Flores, 138
Indricotherium transouralicum, 128
Industrial Revolution, 95-99, 196-97, 208-9; David Brower on, 25, 149; and development of geological science, 26-27; and rising CO_2 levels, 16, 98-99; as widely discussed starting point for Anthropocene, 43, 45, 53, 91
Indus Valley, 167, 175-76, 190
International Chronostratigraphic Chart. *See* geological timescale
International Commission on Stratigraphy (ICS), 3, 64, 189
International Geosphere-Biosphere Programme, 42
International Union for the Conservation of Nature, 37
International Union of Geological Sciences (IUGS), 64, 90
Iranian Plateau, 129, 162, 167, 174, 176, 181
iron, 80, 96, 102, 180, 182, 183

Precambrian time, 86, 98, 112–15
Pueblo cultures, 185
punctuated equilibrium, 29

Quaternary period, 3, 125, 142–43. *See also* Holocene epoch; Pleistocene epoch

Red Sea, 156, 163, 182
Revkin, Andrew, 43
rewilding, 19, 201
rice, 46, 96, 168, 172, 176, 177, 182
Ridley, Matt, 23, 84, 141
Roaring Forties, 19–20
Rocky Mountains, 103
Roelvink, Gerda, 54–55, 82
Romanticism, 27
Rome, 91, 180–81, 186
Ruddiman, William, 46–47, 91, 94
Rudwick, Martin, 27–28
Ryukyu Trench, 30

Sahara, 162, 166, 170, 171–72, 181, 186, 188; desertification of, 174, 175
Sahelanthropus tchadensis, 140
Sato, Makiko, 146, 149, 196
Scandinavia, 153, 157, 172
Science, 51
sea level: early Holocene rise in, 164–65, 169, 170; future, 76, 79–80, 81, 82; pre-Holocene, 21, 116–17, 118, 124, 130, 132, 153; recent rise in, 1, 89
sedentism, 147, 157–60, 164, 168
Shea, John J., 136–38
sheep, 165, 166, 167, 171–72, 175–76, 177, 179
Shetland Islands, 170
Siberia, 39–40, 119, 122, 138, 173
Silk Routes, 163, 181, 182
Silurian period, 118, 125

Skeptical Science, 23
Smail, Daniel Lord, 139
Smith, Bruce, 47–48, 66, 68, 74
Smithfield Foods, 35–36, 205
snowball earth, 29, 113–15
soapberry bug, 79
South Africa, 139, 153
Southern Ocean, 32, 128, 157
South Shetland Islands, 30
South Sudan, 51
Soviet Union, 101, 105, 209
Spain, 91
species relocation, 37–38, 73, 92–93; as stratigraphic marker, 78–79, 81, 93, 98, 102, 142–43, 205
speleothems, 98, 106, 190
spheroidal carbonaceous particles, 103–4
stalagmites, 98, 106, 190
steel, 80, 102, 182
Stoermer, Eugene, 42–43, 44, 48, 49–51, 55
Stoppani, Antonio, 43, 76–77, 81, 204
Storegga Slide, 170
stratigraphy. *See* Anthropocene Working Group; geological timescale
stromatolites, 123
sugar, 92, 96
Sumer, 175
Sundaland, 156
sustainability, 6, 13, 148, 198–200, 202, 208; Paul Crutzen on, 49–50; sustainable development, 56
Sustainocene, 52, 197
Swahili civilization, 183–84, 186

Taiwan, 176, 180
Tambora, Mount, 99
tapirs, 128
Tarantian age, 191
teleconnections, 8–9, 96

Tethys Seaway, 124, 129, 134
Thailand, 166–67, 177
thermohaline circulation, 130–32, 157, 158–59
thrombolites, 123
Tigris, 162, 171, 179
Toba supereruption, 29
trans-Eurasian exchange, 177
Treptichnus pedum, 98
Triassic period, 95, 123–24, 125
trilobites, 11, 117, 126, 145, 193
Tudge, Colin, 24–25, 84, 141
Turner, J. M. W., 99
Tyrannosaurus, 124, 126

Ubaid culture, 170–71
United Kingdom. *See* Britain
United States, 17, 78, 96, 153, 174; nuclear program, 104, 105, 107, 209; twentieth-century history of, 20, 50–51, 100, 101, 197
United States National Grain and Feed Association, 205–6
"Urban Stratum," 81–82
Uruk, 174–75

vegeculture, 173, 177
Venus, 59, 60
Vernadsky, Vladimir, 43

Victoria, Lake, 37–38
Vince, Gaia, 224n4
volcanism, 10, 88, 99, 114, 127; role in mass extinctions, 122–23, 124, 133

Walker, Michael, 190–91
Waters, Colin, 89
Watt, James, 43, 95, 96
wheat, 157, 163, 165, 171–72, 177, 179, 182, 191–92
wilderness, 25, 54, 193
Williams, Mark, 65, 98
Wilson, E. O., 43
woodpecker, black-backed, 19
Wrangel Island, 173, 188
writing, development of, 173, 175

Yeatts, James, 209
Younger Dryas, 158–59, 169

Zagros Mountains, 162, 165, 167
Zalasiewicz, Jan, 66, 77–83, 85, 143, 204; and Anthropocene Working Group, 12, 64–65, 69, 71
Zanclean flood, 129–30, 169
Zeder, Melinda, 47–48, 66, 68, 74
Zhou Enlai, 17
Zimbabwe, 186